prenatal roots of instinctive behavior

SYNTHESIZING

Nature–

Nurture

GILBERT GOTTLIEB

SYNTHESIZING NATURE-NURTURE
Prenatal Roots of Instinctive Behavior

Members of Psychology Laboratory, Dorothea Dix Hospital, at Truby's Lake outside Raleigh, North Carolina (1964): (from left) Jo Ann Winfree, Fred Smith, Anne W. Smith, Gilbert Gottlieb, Patricia Bush, Marvin Simner, Adrian Fountain. (A portion of the author's son Jonathan's head can be seen off Anne Smith's right shoulder.) (Photograph by Nora Lee Willis Gottlieb)

Members of Psychology Laboratory, Dorothea Dix Hospital, Raleigh, North Carolina (1980): (from left) Timothy D. Johnston, David B. Miller, Francisca Jarur, Richard Scoville, Gilbert Gottlieb, Jo Ann Bell, Adrian Fountain, Kathy Bobseine. (Photograph by David B. Miller)

SYNTHESIZING NATURE–NURTURE
Prenatal Roots of Instinctive Behavior

Gilbert Gottlieb
*University of North Carolina
at Chapel Hill*

 LAWRENCE ERLBAUM ASSOCIATES, PUBLISHERS
1997   Mahwah, New Jersey

QL
696
,A52
G67
1997

ACW 0681

Lawrence Erlbaum Associates, Inc., Publishers
10 Industrial Avenue
Mahwah, NJ 07430

Cover design by Nora W. Gottlieb

Library of Congress Cataloging-in-Publication Data

Gottlieb, Gilbert, 1929–
    Synthesizing nature–nurture : prenatal roots of instinctive
behavior / Gilbert Gottlieb.
        p.   cm.
    Includes bibliographical references and index.
    ISBN 0-8058-2548-7 (cloth)   ISBN 0-8058-2870-2 (pbk.)
    1. Mallard—Behavior.   2. Epigenesis.   3. Nature and nurture.
4. Instinct.   I. Title.
QL696.A52G67   1997
598.4'11512—dc21   96-50907
                                                         CIP

10 9 8 7 6 5 4 3 2 1

By way of registering my profound appreciation for their unusual level of dedication, I warmly dedicate this monograph to my full-time research assistants:

Anne W. Smith, 1961–1963
Jo Ann Bell (née Winfree), 1963–1982
Patricia Willis (née Bush), 1963–1965
Mary Evelyn Hale (née Strickland), 1965–1969
Carol S. Ripley, 1965–1970
Mary Catharine Jackson (née Vick), 1970–1978
Kathy Bobseine, 1977–1981
Robert Wiard, 1981–1983
Margaret L. Horton, 1984–1987
Ramona M. Rodriguiz, 1987–1989
Maria V. Collins, 1990–1992
and to the late Marvin Sykes,
machinist *extraordinaire*, 1960–1992.

But nothing is so easy to ignore
as something that does not
yield freely to understanding.

—Giorgio de Santillana

Contents

John M. MacEachran
Memorial Lecture Series

The Department of Psychology at the University of Alberta inaugurated the MacEachran Memorial Lecture Series in 1975 in honor of the late John M. MacEachran. Professor MacEachran was born in Ontario in 1877 and received a PhD in Philosophy from Queen's University in 1905. In 1906 he left for Germany to begin more formal study in psychology, first spending just less than a year in Berlin with Stumpf, and then moving to Leipzig, where he completed a second PhD in 1908 with Wundt as his supervisor. During this period he also spent time in Paris studying under Durkheim and Henri Bergson. With these impressive qualifications the University of Alberta was particularly fortunate in attracting him to its faculty in 1909.

Professor MacEachran's impact has been significant at the university, provincial, and national levels. At the University of Alberta he offered the first courses in psychology and subsequently served as Head of the Department of Philosophy and Psychology and Provost of the University until his retirement in 1945. It was largely owing to his activities and example that several areas of academic study were established on a firm and enduring basis. In addition to playing a major role in establishing the Faculties of Medicine, Education, and Law in the Province, Professor MacEachran was also instrumental in the formative stages of the Mental Health Movement in Alberta. At a national level, he was one of the founders of the Canadian Psychological Association and also became its first Honorary President in 1939. John M. MacEachran was indeed one of the pioneers in the development of psychology in Canada.

Perhaps the most significant aspect of the MacEachran Memorial

Lecture Series has been the continuing agreement that the Department of Psychology at the University of Alberta has with Lawrence Erlbaum Associates, Publishers, Inc., for the publication of each lecture series. The following is a list of the Invited Speakers and the titles of their published lectures:

1975   Frank A. Geldard (Princeton University)
        Sensory Saltation: Metastability in the Perceptual World

1976   Benton J. Underwood (Northwestern University)
        Temporal Codes for Memories: Issues and Problems

1977   David Elkind (Rochester University)
        The Child's Reality: Three Developmental Themes

1978   Harold Kelly (University of California, Los Angeles)
        Personal Relationships: Their Structures and Processes

1979   Robert Rescorla (Yale University)
        Pavlovian Second-Order Conditioning: Studies in Associative Learning

1980   Mortimer Mishkin (NIMH-Bethesda)
        Cognitive Circuits (unpublished)

1981   James Greeno (University of Pittsburgh)
        Current Cognitive Theory in Problem Solving (unpublished)

1982   William Uttal (University of Michigan)
        Visual Form Detection in 3-Dimensional Space

1983   Jean Mandler (University of California, San Diego)
        Stories, Scripts, and Scenes: Aspects of Schema Theory

1984   George Collier and Carolyn Rovee-Collier (Rutgers University)
        Learning and Motivation: Function and Mechanism (unpublished)

1985   Alice Eagly (Purdue University)
        Sex Differences in Social Behavior: A Social-Role Interpretation

1986   Karl Pribram (Stanford University)
        Brain and Perception: Holonomy and Structure in Figural Processing

1987   Abram Amsel (University of Texas at Austin)
Behaviorism, Neobehaviorism, and Cognitivism in Learning Theory: Historical and Contemporary Perspectives

1988   Robert S. Siegler and Eric Jenkins (Carnegie Mellon University)
How Children Discover New Strategies

1989   Robert Efron (University of California, Martinez)
The Decline and Fall of Hemispheric Specialization

1990   Philip N. Johnson-Laird (Princeton University)
Human and Machine Thinking

1991   Timothy A. Salthouse (Georgia Institute of Technology)
Mechanisms of Age-Cognition Relations in Adulthood

1992   Scott Paris (University of Michigan)
Authentic Assessment of Children's Literacy and Learning

1993   Bryan Kolb (University of Lethbridge)
Brain Plasticity and Behavior

1994   Max Coltheart (Maquarie University). *Our Mental Lexicon: Empirical Evidence of the Modularity of Mind (unpublished)*

1995   Norbert Schwarz (University of Michigan)
Cognition and Communication: Judgmental Biases, Research Methods, and the Logic of Conversation

1996   Gilbert Gottlieb (University of North Carolina at Chapel Hill)
Synthesizing Nature–Nurture: Prenatal Roots of Instinctive Behavior

Eugene C. Lechelt, Coordinator
MacEachran Memorial Lecture Series

Sponsored by The Department of Psychology, The University of Alberta, in memory of John M. MacEachran, pioneer in Canadian psychology.

Preface

This small book describes our present understanding of how nature and nurture collaborate to bring about normal and abnormal behavioral and psychological development. Having the species-normal complement of genes does not guarantee that the outcome of an individual's development will be normal. That is because the expression of our heredity takes place in the context of other influences that act above the level of genes and determine whether or not genes will express themselves. In other words, genes do not make behavior happen, even though behavior won't happen without them.

To document this point of view, I tell the story of my own research on how the expression of instinctive behavior in newborn animals is regulated in a nonobvious way by their prenatal experience. I chose to do this research with bird species because the egg and the embryo develop outside the body of the mother and are more amenable to experimental study than in mammals, in which the fetus develops inside the body of the mother.

Because this research has already been published in technical journals, I decided to take a less technical, more conversational approach to telling the story in the hope that readers outside the field would be encouraged to follow a meandering trail as the experiments took me into mostly uncharted territory between biology and psychology. I hope you enjoy sharing this journey as much as I enjoyed making the trip.

ACKNOWLEDGMENTS

I received a very high level of stable financial support for my research program through funds allocated by the North Carolina Department of

Mental Health to its Division of Research, of which I was a member from 1961 to 1982, at Dorothea Dix Hospital in Raleigh, North Carolina. Other financial support came from the National Institute of Child Health and Human Development (1962–1985), National Science Foundation (1985–1988), and the National Institute of Mental Health (1988 to present).

Generous laboratory facilities were made available to me at Dorothea Dix Hospital (including an animal behavior field station) from 1959 to 1982, and at the University of North Carolina at Greensboro (1982–1995). In September 1995, I became a Research Professor of Psychology in the Center for Developmental Science, University of North Carolina at Chapel Hill, where I am a member of the Carolina Consortium on Human Development.

Because this small book covers virtually my entire research career to date, there are many other people to thank in addition to those mentioned on the dedication page. Carter Doran, W. "Mac" Euliss, David H. Gottlieb, Linda Green, Mary Haskett, Jeffrey Hill, Laura Hyatt, Francisca Jarur, William Lynn, Kenneth Ripley, and Carolyn Schmonsees served as part-time research assistants.

David B. Miller completed the field research on the maternal calls of mallard and wood ducks that I and my wife, Nora W. Gottlieb, initiated in 1961–63.

Fred Davis III improved the sound-delivery system of the test apparatus, and designed and constructed a frequency chronogram and an FM pulse-shaping network for making the synthetic wood duck maternal calls (chapter 5). Further in connection with chapter 5, James Ward's finely tuned musician's ear convinced me that at least some humans are as aurally capable as wood ducklings (I am not) in distinguishing ascending-descending from descending frequency modulations.

Egg collection in the field to provide wood ducklings for our experiments in the laboratory often involved grappling with 5-ft-long black snakes, raccoons, opossums, and tenacious wood duck hens, all of which can be found in the wood duck nest boxes that I installed at our farm, the Dorothea Dix Animal Behavior Field Station, and, with the assistance of J. V. Clifton, on my late neighbor Truby Upchurch's lake. For their some-times heroic egg collection efforts, in order of appearance, I thank Eugene Barrett, Roger Montague, William Lynn, Randall Pittman, Lawrence Miller, Jeffrey Hill, Marc S. Gottlieb, W. Thomas Tomlinson, and Matthew and Justin Schneiderman. Also thanks to my sons, Jonathan, David, Aaron, and Marc Gottlieb, for helping at the field station at our home and for their shouts of encouragement during the wood duck egg collection adventures.

Among the predoctoral students who worked in my laboratory are the following: Lincoln Gray (junior high school science fair project), now

Professor of Otolaryngology and Director of Research, University of Texas Medical Center; Marvin L. Simner (master's thesis), now Professor of Psychology, University of Western Ontario, Canada; Patricia C. Arrowood (master's thesis); Marieta B. Heaton (doctoral thesis), now Professor of Neuroscience, University of Florida Medical School; Richard Scoville (doctoral thesis), resigned in third year of tenure-track position at Virginia Polytechnic and State University to go into computer work; Barbara Lawrence (honors thesis), now Associate Dean at a college in the Midwest; Gang Wang (master's thesis), now doctoral candidate in neuroscience at Tulane University; W. Thomas Tomlinson and Peter Radell (doctoral students); Mary Elizabeth Lovin (master's thesis); and Cheryl Ann Sexton (doctoral thesis), now postdoctoral fellow, Carolina Consortium on Human Development, Center for Developmental Science, University of North Carolina at Chapel Hill.

Finally, I was most fortunate to have pass through my laboratory an intellectually stimulating group of postdoctoral research associates: Marshall Harth (1970–1972), now Professor of Psychology, Arapahoe College, New Jersey; William Lippe (1972–1975), now Research Associate, Department of Otolaryngology and Neck Surgery, University of Washington School of Medicine; David B. Miller (1974–1980), now Professor of Psychology, University of Connecticut; Timothy D. Johnson (1978–1983), now Professor of Psychology, University of North Carolina at Greensboro; Robert Lickliter (1983–1986), now Professor of Psychology, Virginia Polytechnic Institute and State University; Charles F. Blaich (1986–1987), now Associate Professor of Psychology, Wabash College, Indiana; Antoinette Dyer (1987–1988), now Assistant Professor of Psychology, Radford University; and Lubov Dmitrieva (1990–1992), former Research Associate, USSR Academy of Sciences, Institute of Higher Nervous Activity and Neurophysiology, in the late and sorely missed Sergei Khayutin's Laboratory of Behavioral Ontogeny, Moscow.

Ramona M. Rodriguiz, a former research assistant and now completing her doctorate at the University of North Carolina, Chapel Hill, has provided bibliographic and other aid, for which I am most grateful.

The final portion of this monograph was completed while I was a Visiting Fellow at The Neurosciences Institute, San Diego, California, from February through April 1996. My thanks to Gerald Edelman and W. Einar Gall for their generous hospitality during my visit. Lisa Murphy's cheerful diligence in helping to prepare the manuscript for the printer is gratefully acknowledged.

I appreciate the permission of the American Psychological Association to reproduce here (chapter 6) portions of two of my articles that appeared in *Developmental Psychology,* 1991, *27*, 4–13 and 35–39: Experiential canalization of behavioral development: Theory and results. The historical material in chapter 8 was also reviewed in Gottlieb (1996). A portion of chapter 9 was modified from chapter 14 in Gottlieb (1992).

I also appreciate the permission to reproduce the work of others: Edwin Rubel and the Neurosciences Research Foundation (Fig. 4.5), Ursula Bellugi (Fig. 8.1), University of Chicago (Fig. 9.2), George L. G. Miklos (Table 9.1), R. A. Raff and T. C. Kaufman (Fig. 9.1), E. J. Brill (Fig. 9.3), and J. S. Wyles and J. G. Kunkel (Fig. 9.4).

The author's research and scholarship are currently supported by grant MH-52429 from the National Institute of Mental Health.

1

Formulating Probabilistic Epigenesis

I was very fortunate early in my postgraduate career to have stumbled onto a research finding that kept me gainfully employed for 35 years. I recapture the high points of that intellectual adventure in the chapters to follow, but first I would like to describe the personal context out of which the research program actually developed.

Perhaps the reader should be forewarned that this monograph is not standard fare, because in each chapter I include relevant background considerations and autobiographical details that almost never get into journal articles or scientific monographs, but are nonetheless pertinent to an understanding of the intellectual path taken by the investigator and, thus, to an understanding of why the research program developed in the way that it did. Everyone knows that there is a significant personal side to science, but it is only rarely made public, largely because the tradition of scientific reporting discourages the recounting of personal considerations. Otherwise, this is the story of how the instinctive behavior of ducklings is created out of their experience in the egg.

GETTING STARTED: UNDERGRADUATE BEGINNINGS

I became a serious university student only after a 6-year hiatus in the real world between my sophomore and junior years. The most influential experience was service in the U. S. Occupation Forces in Europe after World War II. My job brought me into contact with displaced persons, and I was struck by the immense individual differences in coping under unusually stressful circumstances. I wanted very much to understand these individual differences and resolved to enter the field of psychology after my discharge. (It was, of course, naive to think that psychology could supply the answer, but it was an appropriate place to at least get oriented to the problem.)

After being discharged from the Army at the age of 24, I entered the University of Miami in my junior year in January of 1954. Although I was quite interested in the formal courses I took, I was thirsting for something beyond what I was being exposed to in the classroom, and suspected the professors were not telling all they knew. I spent a lot of time reading, not quite randomly, in the university library. I was drawn to books about evolution, books about embryology, late 19th and early 20th century theosophy, Freud's writings on psychoanalysis, and the *Journal of Experimental Psychology*. Although I was doing quite well in class, I understood little of what I encountered in my self-directed reading program and actively misunderstood what I encountered in the *Journal of Experimental Psychology*. My misunderstanding of the contents of *JEP* supported my ill-founded belief that the psychology professors were indeed holding back the choicest intellectual morsels in their classroom lectures. Why, in the pages of every issue of *JEP*, psychologists were reporting the outcomes of their experiments with the unconscious and conscious minds, not only of humans, but of rats! Fortunately, I kept these insights to myself, for such was my understanding that I thought the Pavlovian notations UCS (for unconditioned stimulus) and UCR (unconditioned response) were abbreviations for the unconscious and CS (conditioned stimulus) and CR (conditioned response) were abbreviations for the conscious. No wonder I did not understand *anything* I read in the *JEP*.

Along with my courses in psychology, I took many courses in philosophy as well as intellectual history (the history of ideas), because, while I was obviously quite intellectually naive, I did know the question I was interested in. I wanted to know the appropriate intellectual framework for coming to an understanding of events and things in the world. I dimly grasped that development (a person's history of experiences) was central to this understanding, so that is why I read Freud (on my own). Otherwise, my theoretical understanding was guided largely by an intuitive feeling of rightness. Alfred North Whitehead's (1929) notion of "the process character of reality" seemed right to me as an undergraduate, as it still does today. Among other things, development signifies change.

In my senior year I had the good fortune, at the suggestion of an English professor, Richard Royce, to home in on three books that, at the time, completely satisfied my search for an appropriate metatheoretical framework for gathering valid insights into events and things in the world. My personal bibles were Dewey and Bentley's (1949) *Knowing and the Known*,

Egon Brunswik's (1952) monograph on *The Conceptual Framework of Psychology*, and Harry Stack Sullivan's (1953) *The Interpersonal Theory of Psychiatry*.

Dewey and Bentley described in a historical way how science has proceeded from self-actional explanatory frameworks (animism, self-acting souls or minds, instincts) to interactional frameworks (primarily Newton's mechanics), and, finally, to transactional frameworks (their term for seeing events and things in their full historical, cultural, evolutionary setting; a social and biological science derivative in harmony with field theory in physics). Whereas in the classical interactional explanatory framework the interacting bodies remain fundamentally unchanged (e.g., billiard balls only change their direction after they collide), in the transactional framework the components themselves become transformed (e.g., our present-day understanding of the consequences of infant–caretaker or peer–peer social encounters). At the University of Miami, I became such a vocal devotee of the transactional point of view that I raised these issues in and outside of class and was disappointed to find that, by and large, my psychology professors were ignorant of Dewey and Bentley's work, and, furthermore, at least one of them did not welcome this way of thinking about scientific explanation in psychology. I now recognize that I probably behaved as an arrogant nuisance in my eagerness to press my new-found "knowledge" on anyone within hearing distance.

Brunswik's (1952) treatise likewise gave a historical account of psychology's groping for appropriate methods and theories, eventually converging on a molar, nonreductive behaviorism (à la E. C. Tolman, for example) and a probabilistic functionalism, both in tune with organismic or biological field theory or what we would today call a developmental systems outlook (more on this later).[1]

Sullivan's (1953) interpersonal theory of psychiatry moved beyond Freud's primarily intrapsychic (within the mind) theory to the observable social events that influence our personality development beginning in infancy, with significant psychobiological changes occurring in each further developmental *era* or *epoch*, as he called them: childhood, the juvenile era, preadolescence, early and late adolescence. Here was a theory of the development of the self that appealed to me for its practical utility (I was planning to be a clinical psychologist) and its amenability to a scientific test

---

[1]Should advanced undergraduate or graduate students decide to pursue the aforementioned books, I would urge them not to be discouraged by the intellectual density of these tomes. To this day I still read these works with benefit and, although most developmental psychologists will claim some acquaintance with the notion of transaction, they will most likely not have drunk at the source. The same can be said for Brunswik's undercited masterpiece. Sullivan's book, edited and published posthumously by his close colleagues, has enjoyed a wide readership.

(I was also planning to be an experimental psychologist as well, both of which I subsequently achieved at Duke University).

With respect to my early attraction to Sullivan's writings, it is interesting to note that my research and theorizing are in the tradition of what has been called developmental psychobiology. Clarence Luther Herrick (1858–1904) christened the field in the late 1800s, identifying it as the comparative study of the nervous system, behavior, and psychology from the standpoints of embryology, anatomy, physiology, and eventually philosophy (Gottlieb, 1987a). Adolf Meyer, the first Dean of American Psychiatry at Johns Hopkins University, adopted and enlarged upon Herrick's holistic approach, and is sometimes referred to as the father of the distinctively American school of psychobiological psychiatry. Harry Stack Sullivan readily absorbed Meyer's congenial ideas when, in 1922, he took a position at the Sheppard and Enoch Pratt Hospital, a private psychiatric clinic outside of Baltimore, Maryland, where Meyer headed the Henry Phipps Psychiatric Clinic at Johns Hopkins (Perry, 1982).

GRADUATE SCHOOL MUSINGS

In graduate school at Duke University my interests in development and evolution (very generally speaking) became more congealed, but yet another intuitive understanding of "rightness" intruded itself, one that I knew once again was correct but I could not say why in an intellectually defensible way. Obviously not really having understood fully what I had read in Brunswik about the desirability of the "representative design of experiments," I decided it was essential in psychology to study real-world things and events, and not use completely artificial stimuli or stimulus objects or non-species-typical behavioral responses, even if one were going to use laboratory procedures. This is, of course, precisely what Brunswik had in mind, although I did not grasp his way of putting it at the time.

Finally, on the developmental side, I began to think that the study of infants was essential, with the aim of pursuing the prenatal antecedents, if that were to prove possible. Although I don't recall that Sullivan had anything to say about the prenatal period, for intuitive reasons I cannot explain, I became convinced of the importance of prenatal experience for the adaptive (or maladaptive) behavior of the infant. (Helen Swick Perry, in her biography of Sullivan [1982], mentions that Sullivan lectured on the prenatal period to attendants at his clinic but no details have come down to us. He began his *Interpersonal Theory of Psychiatry* [1953] with infancy.)

It was around this time that a fellow graduate student, Ann Lodge, called my attention to an article in *Scientific American* on imprinting by Eckhard Hess (1958). Thus it was that I happily blundered into the developmental study of species identification in ducklings. That kind of research involved a developmental study of infants, with the remote possibility (for a non-biologically trained person) of determining possible overt embryological precursors in the embryo–egg behavioral or experiential situation. The following-response was something ducklings did in nature, and, apparently, the ducklings became "imprinted" to their mother in nature, as one read in Konrad Lorenz's (1937) very famous paper. And the laboratory research of psychologists such as Eckhard Hess and Julian Jaynes seemed to support Lorenz's contention about the swiftness of the learning, its limitation to a brief critical period early in development, and its retention for hours and days (if not longer).

During the oral defense of my doctoral dissertation on imprinting in mallard ducklings, one of the inquisitors noted that I had indeed demonstrated something like a critical or sensitive period, but even at the height of that period not all of the ducklings followed the model of a mallard hen. He asked if they would not all follow the hen in nature, and, although I agreed with the implication, I realized I did not really know the answer to that question or actually what went on in nature, because Lorenz's main observations (and those of his predecessors Oskar Heinroth and Douglas Spalding) were made on waterfowl and chicks hatched in incubators and imprinted to human beings.

TAKING TO THE FIELD: EARLY POSTGRADUATE RESEARCH

While I was writing my doctoral dissertation, in November 1959, I took a job as a clinical psychologist at Dorothea Dix Hospital in Raleigh, North Carolina, a state-run psychiatric facility. I took the job because the administrators allowed me to set up an animal behavior laboratory at the hospital and granted me one day a week to work in my laboratory. In 1959, a Division of Research was established at Dorothea Dix Hospital by the North Carolina Department of Mental Health. In January of 1961 I was hired as the first basic research scientist in the division, and that afforded me the time to make observations of "imprinting" in nature. With the necessary assistance of my wife, Nora Lee Willis Gottlieb, and my friend, Gus Martin, and with the cooperation of Eugene Hester, of Wendell, North Carolina, and John Whalen, of Bath, North Carolina, I was able to photograph and tape-record the events before, during, and after hatching in two species of waterfowl, wood ducks and mallard ducks. (Figures 1.1 and 1.2 show the exodus from the nest in a ground-nesting mallard and a hole-nesting wood duck.)

Until the time of our naturalistic observations in 1961–1963, imprinting

FIG. 1.1. Ground-nesting mallard hens readily take to artificial nest boxes placed near the water. (1) Before leading her hatchlings off the nest, 1 to 2 days after they have hatched, the hen makes a reconnaissance of the area. (2) If she detects no intruders, the hen utters her maternal assembly call and leads her young from the nest. The family stays together until the young are able to fly (about 8 weeks) at which time the family disbands.

had been understood as primarily, if not solely, a visual phenomenon: The young birds showed evidence of learning the visual characteristics of the objects that they followed in the laboratory. Peter Klopfer (1959) had shown that hole-nesting wood ducklings could learn the characteristics of artificial sounds he played to them but ground-nesting species such as mallards did not learn the artificial auditory signals, and all of the other laboratory research was with ground-nesting species.

Our observations of duck families in nature indicated that the hen's voice (assembly call) played a prominent role in the maternal–infant relationship at all stages. In both hole-nesting and ground-nesting species, the hen began softly uttering her assembly call during the pipping stage before her ducklings hatched. In both species, she used this same call to entice the young out of the nest after they hatched, and she continued to periodically utter the call for the next 6–8 weeks as her brood grew to fledging size, at which time the family disbanded. (Adult female mallards and wood ducks utter the call upon encountering food, and that brings their mate promptly to their side, so the assembly call retains its approach-evoking quality throughout life. Only females utter the call.)

The basic idea behind imprinting was that the young duckling did not hatch with a picture of its species inside its head, and that it acquired this perception through exposure to the visual characteristics of its mother or other caretaker (a human, in the case of Konrad Lorenz and the dozens of others of us who have misdirected the following-response of young ducklings over the years). In nature, of course, the hen would be the first adult social object that ducklings would ordinarily encounter, so this process was a good example of evolution or natural selection being able to use a recurring developmental organismic experience, rather than instinct, to propagate species identification or recognition from generation to generation. I thought the same would be true for the ducklings' ability to identify the maternal assembly call of their own species, each species' maternal call being distinctive and acoustically different from the assembly calls of other species. Because the ducklings, even the embryos, were inevitably exposed to their mother's assembly call both before and after hatching, they could learn the species-specific maternal call anew in each generation. I was in for a surprise that shaped my entire postgraduate research career.

FLEDGLING DEVELOPMENTAL PSYCHOBIOLOGIST FACED WITH INSTINCT

Armed with the tape recordings of a number of maternal assembly calls of various species, I did what I thought to be an obvious experiment to document the necessary learning or imprinting of the call by the young: I tested wood ducklings and mallard ducklings hatched in incubators and assumed that they would be unable to identify the maternal call of their species in the absence of prior exposure to the call. In the most basic test, I placed incubator-hatched wood and mallard ducklings, singly, in a circular arena in which one speaker broadcast a wood duck assembly call and another speaker broadcast the mallard assembly call, at opposing 90-degree angles from the point at which the duckling was put in the arena (Fig. 1.3). Almost without exception, the wood ducklings approached the speaker broadcasting the wood duck call, and the mallard ducklings approached the speaker broadcasting the mallard call. As a further check, we tested incubator-hatched junglefowl and domestic chicks to their maternal assembly call with the same result. And we added choice tests for the mallards and wood ducklings, which involved their species' maternal call versus chicken, mandarin, and pintail maternal calls, always with the same result: The ducklings approached the speaker broadcasting the maternal call of their own species even though they had not heard it beforehand.

I did not at first appreciate the magnificent opportunity for a thorough-going developmental analysis that nature had placed before me. Rather, under the influence of the imprinting ethos, I tried to alter the ducklings' species-specific auditory preference by having them follow a stuffed hen emitting the maternal call of another species (Gottlieb, 1965). This effort was only partially successful, and only served, eventually, to convince me of the strength of the species-specific auditory preference. The experimental results were also forcing me to relinquish my belief in visual imprinting as

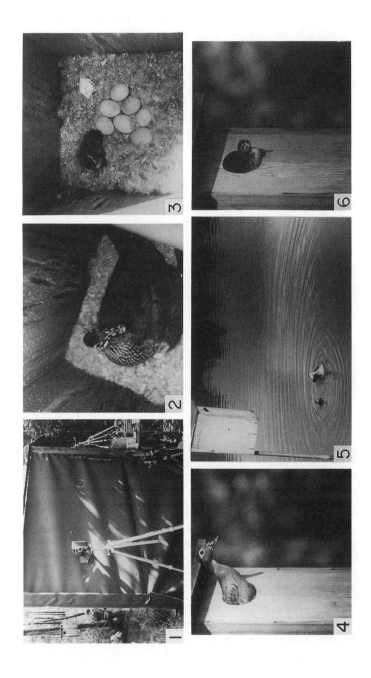

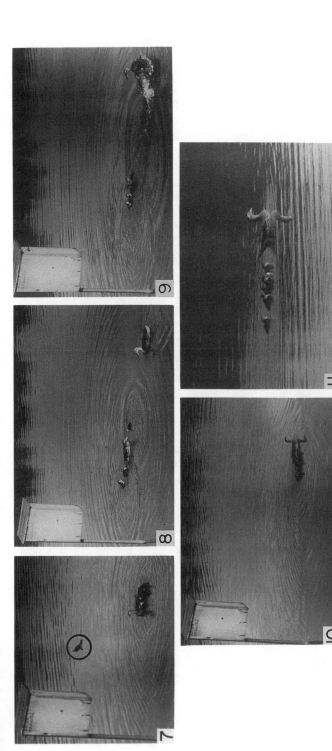

FIG. 1.2. Exodus of hole-nesting wood ducklings. (1) Camera and tape recorder are obscured from ducks by the blind. Microphone is hidden under nest. (2) Hen drops "protective" wing over clutch as intruder lifts lid of hole-nest. (3) The hen was deliberately flushed. The young hatch over a rather protracted period (as much as 18 hr in some rare cases). (4) Just prior to the exodus, 1–2 days after the young have hatched, the hen appears at the exit and makes a reconnaissance of the area before (5) dropping below the nestbox and calling the young out. (6) The young climb up the wall inside the nestbox, pause momentarily at the exit, and (7) then leap to the water below. (8) Should another duck intrude during the exodus, (9) the hen attacks. (10) Within about 4 min all the ducklings are out of the nestbox. (11) The hen leads her brood to an area of thick cover in a nearby swamp where the young grow to flight stage (about 8 weeks). At that time the family disbands.

9

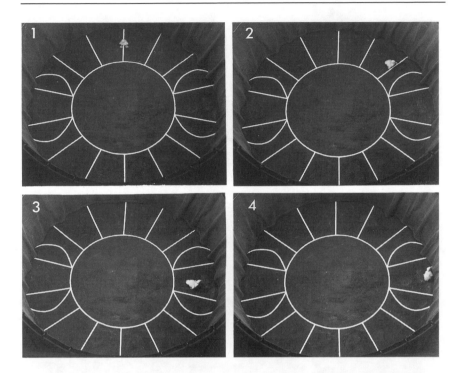

FIG. 1.3. (a) Simultaneous auditory choice test. (1) Duckling placed equidistant between two speakers (not visible), in front of which are painted elliptical approach areas. (2) Duckling on its way toward approach area. (3) Duckling in approach area. (4) Duckling snuggling to curtain and orienting to nonvisible speaker broadcasting maternal call of its species.

the primary mechanism for species identification. For example, with the able assistance of the laboratory's machinist, Marvin Sykes, we devised an apparatus in which we could test incubator-hatched ducklings and chicks for their preference between a moving, visual replica of the hen of their species versus an unseen moving sound source broadcasting the maternal call of their species (Fig. 1.4). All of the ducklings, both the hole-nesting wood ducklings and the ground-nesting mallard ducklings, as well as the ground-nesting chicks, followed the maternal call, even though they had not previously been exposed to the call.

INITIAL FORMULATION OF PREDETERMINED AND PROBABILISTIC EPIGENESIS

During the period under discussion (1962–1965), two notable things happened. One, I was fortunate enough to convince Zing-Yang Kuo to come to

FIG. 1.3. (b) Research assistant Jo Ann Bell (née Winfree) turning off stereophonic tape player at end of 5-min auditory choice test.

my laboratory from Hong Kong for 6 months to teach me his techniques of behavioral embryology (Fig. 1.5), and two, I was working very hard on a new way to express the nature-nurture dichotomy, one that not only would make the respective concepts subject to experimental test, but would clarify the different developmental pathways involved. The stimulus to commit to a written version of this theoretical work was an invitation by Ethel Tobach to prepare a chapter for a *Festschrift* for T. C., Schneirla, who was planning to retire from the Department of Animal Behavior at the American Museum of Natural History and join our research group at

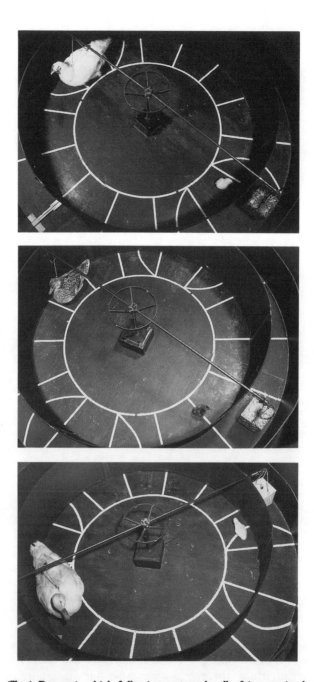

FIG. 1.4. (Top) Domestic chick following maternal call of its species broadcast from a concealed moving sound source in preference to silent, moving, visual maternal replica in an auditory versus visual choice test. (Middle) Mallard duckling following maternal call of its species in an auditory vs. visual choice test. (Bottom) Domestic mallard (Peking) duckling doing the same.

FIG. 1.5. When Z.-Y. Kuo came to the States from Hong Kong in August of 1963, I met him in Washington, D.C., where we were both scheduled to give talks at the International Congress of Zoology meeting. Kuo had been inactive for quite a while, but his theories were kept alive (described later) not only in a positive way by T. C. Schneirla and D. S. Lehrman, but also in a negative way by certain ethologists such as W. H. Thorpe, who most likely presumed Kuo to be deceased. (Their attacks were sometimes nasty.) Kuo had a quaint sense of humor and asked that as many leading ethologists (both friends and foes) as were attending the meeting be invited to a dinner to be hosted by him. Clockwise from the far left, N. Tinbergen (Oxford University), author, J. P. Kruijt (University of Groningen), D. E. Davis (Pennsylvania State University), Dr. Sherman (American Psychological Association), J. T. Emlen (University of Wisconsin at Madison), G. P. Baerends (Groningen), Kuo, E. Fabricius (University of Stockholm), M. W. Schein (Pennsylvania State University), J. P. Scott (Bowling Green State University), W. H. Thorpe (Cambridge University), and our hostess, a friend of Dr. Kuo's, whose name I no longer recall.

Dorothea Dix Hospital.[2] Having been influenced and encouraged by Dr. Schneirla's writings on instinctive behavior, I hoped my chapter would please him.

---

[2]As an aside, although I was most pleased and honored by the invitation to contribute to Dr. Schneirla's *Festschrift*—I had only recently made his acquaintance—I was frustrated that my chapter, completed as requested in 1965, did not appear until 1970, because Schneirla's foremost student, Daniel S. Lehrman, a major contributor, was slow in getting his chapter written. It could not have been otherwise—Lehrman was forever woefully overcommitted and the book, *Development and Evolution of Behavior*, could not be published without his contribution. But this was the first such chapter I had written and I wanted the world to see it on time. It was sadder that Dr. Schneirla died before the book appeared.

FIG. 1.6. Zing-Yang Kuo supervising the author in the making of a Kuo Observation Window in the shell of a duck egg (Dorothea Dix Hospital, Raleigh, NC, fall 1963).

The chapter appeared in 1970 and was titled "Conceptions of Prenatal Behavior"; it covered the research and thinking of a small handful of biologists and two psychologists (Zing-Yang Kuo and Leonard Carmichael) who took a behavioral approach to embryology: behavioral embryology (Fig. 1.6).

I began the article by giving a definition of epigenesis that was consonant with the content of Needham's history of embryology (1959) up to 1900 and then briefly described the two points of view the reader would encounter: predetermined epigenesis (I think this is Needham's term, used once in his book to describe the "modern view," but I am now unable to locate it) and probabilistic epigenesis (my term, though owing something to my reading of Brunswik):

> In the classical usage of the term, all present-day theories of prenatal behavioral development can be characterized as epigenetic. [Schneirla and Lehrman had accused the classical ethologists of being preformistic in their concept of instinct. While that captured the flavor of ethologists' thinking about instinct, instinctive behavior was nonetheless an epigenetic (i.e., an emergent) phenomenon as the term epigenesis was understood in Needham's historical account.] This term denotes the fact that patterns of activity and sensitivity are not immediately evident in the initial stages of embryonic development and that the various behavioral capabilities of the organism became manifest only during the course of development. However, major disagreement exists with regard to the fundamental character of the

epigenesis of behavior. One viewpoint holds that behavioral epigenesis is predetermined by invariant organic factors of growth and differentiation (particularly neural maturation), and the other main viewpoint holds that the sequence and outcome of prenatal behavior is probabilistically determined by the critical operation of various endogenous and exogenous stimulative events. (Gottlieb, 1970, p. 111)

I explicitly avoided the notions of genetic determination versus environmental determination in characterizing predetermined and probabilistic epigenesis, because, to my mind, (a) these are not developmental conceptions and (b) genetic activity is involved in both probabilistic and predetermined epigenesis, so that was not a decisive difference between the two viewpoints.

My reading of Kuo, Myrtle McGraw, Schneirla, and others on the probabilistic side led me to conclude that they were seeing the structure–function relationship differently than the predeterministic camp. Consequently, I proposed a bidirectional (reciprocal) relationship between structure ↔ function as a defining feature of probabilistic epigenesis and a unidirectional structure → function relationship as the cardinal feature of the predetermined view. Up to this point, behavioral embryology had been based almost exclusively on descriptive-correlational studies and I wanted very much to formulate ideas that could make the difference between the two viewpoints experimentally testable.[3]

At the end of the chapter, I was able to conclude:

The single most important issue dividing the two theoretical camps is the role of endogenous and exogenous stimulation in behavioral, neuroanatomical, and musculoskeletal development. At the level of functional anatomy, there are a few recent prenatal experiments that suggest that relevant sensory and musculoskeletal stimulation may be essential to the initiation and maintenance of normal (typically observed) maturational changes. These and other experiments with neonates would seem to indicate a particularly significant shift in our conception of the structure–function relationship from a unidirectional one (structure → function) to a bidirectional one (structure ↔ function). This shift is consonant with a probabilistic conception of epigenesis and raises an important question for future resolution at the prenatal level, namely, the degree to which stimulation or activity merely fosters or enhances development and the degree to which (or areas or stages in which) stimulation or activity channels future development. (Gottlieb, 1970, p. 134)

In this paragraph, I was groping for ways to say that stimulation or activity could function in various ways: They could channel or induce

---

[3]I was very pleased that Dr. Kuo, who had read the draft of my manuscript before it was published, saw fit to mention favorably my concept of bidirectionality in the book he was preparing, which appeared before my chapter (Kuo, 1967).

development, they could enhance development, they could maintain development. Later, I would formalize these notions. Also, I was pleased that it had proven possible to write about "the ancient dualism of nature and nurture" without ever once using the terms learning, instinct, or innateness, but still being able to do justice to the issues at hand. This showed (at least to me) that it truly was developmental questions—different views about developmental pathways—that were at the heart of the nature–nurture debate, and that these were experimentally resolvable if placed in an appropriate developmental conceptual framework.

2

Preliminaries to an Explicit Experimental Demonstration of Probabilistic Epigenesis

Mallard ducklings, wood ducklings, junglefowl, and domestic chicks could identify the maternal assembly call of their own species without prior exposure to it. In 1966, I wrote:

> From an evolutionary point of view, "instinctive" recognition of species in chicks and ducklings . . . can be explained at the population or species level by the operation of natural selection. However, the principle of natural selection does not explain . . . how the ability of chicks and ducklings to identify the general characteristics of their own species arises during prenatal and postnatal development of the individual. (Gottlieb, 1966, p. 282)

That is a developmental question.[1]

Before being tested with the maternal calls, the only vocal–auditory experience the birds had was exposure to their own vocalizations. Even though the neonatal and maternal calls sound very different to a human

---

[1]Although it is now fairly widely understood that natural selection works on already developed outcomes (so it can not have caused those outcomes and, thus, sheds no light on their development), at the time (1966) the point was not widely appreciated. In his long-awaited 1970 chapter for the Schneirla *Festschrift*, Daniel Lehrman made that point in a way that made it understandable to many readers, so he is often given credit for that insight in the behavioral science literature. Years later, I ran across Mivart's (1871) developmental critique of Darwin's use of natural selection to explain evolution, and was able, finally, to give credit to the person to whom it was most surely due (Gottlieb, 1992, chapter 5). Although some colleagues might think that I am overgenerous in stating Mivart's priority, I would much prefer to err in that direction than in the other direction. As George Steiner has remarked somewhere, genuine scholarship involves not only recognition of sources but generosity to intellectual precedence.

observer, perhaps there is some acoustic correspondence that is not evident to a human observer. Because both chicks and ducklings begin vocalizing in the egg several days before hatching, they become amply familiar with the sound of their own and sibling voices. Because young chicks and ducklings approach and follow familiar sources of stimulation, perhaps the maternal call is attractive because it shares acoustic features in common with the chicks' and ducklings' own vocalizations. Learning theorists have a name for this process: *stimulus generalization*. In stimulus generalization, the animal always gives its strongest response to the original stimulus with which it was trained and progressively weaker responses to physically graded stimuli that resemble the training stimulus less and less. If the embryonic or neonatal vocalizations are the training calls and the maternal calls derive their attractiveness from the training calls, in a simultaneous auditory choice test, the chicks and ducklings would prefer to approach recordings of their own vocalizations over recordings of their respective maternal calls.[2]

I was exposed to the thinking of premier learning theorists as a graduate student at Duke. Much as I admired the precision and clarity of learning theory, I never thought that exposure to it would pay off in my own work. However, in the introduction to my 1966 article, I did introduce a caveat: "For purposes of conceptual clarity, it is important to note that prenatal auditory stimulation involving the neonatal call may indeed contribute to the neonate's responsiveness to the maternal call, but that the mode of its contribution may not be via some known phenomenon of learning such as stimulus generalization" (Gottlieb, 1966, p. 283). I wanted to be clear that even if stimulus generalization did not pan out, I was not going to give up on pursuing the possible contribution of the perinatal vocalizations to the birds' ability to identify the maternal call of their species.

The results were clear. Although chicks and ducklings preferred their own vocalizations over the maternal calls of other species, they preferred

---

[2]To keep to the main point of this essay, I am postponing to chapter 4 a description of the research that involved testing the chick and duck embryos' behavioral and physiological responsiveness to the various maternal calls. We had to devise new methods to be able to objectively record the behavior and physiology of the embryo, because behavioral embryology was done solely by visual observation up to this point. I felt it was desirable to put the observations on an objective footing in order to resolve and avoid disputes that could result from the "eyeball method." In launching this research after Dr. Kuo's visit in 1963, I was fortunate to have the excellent services of Patricia Bush Willis, an intrepid research technician, who was also willing and able to work at the 100°F necessary to carry out the initial experiments with embryos in the early 1960s. Lincoln Gray and Evelyn Hale (née Strickland), exceptionally qualified high school students at the time, also ably assisted with these early experiments on chick and duck embryos (Gottlieb, 1965). Eventually, we were able to pipe in cool air to the first embryonic testing room from a remote source so that the sound of the air conditioner could not be heard by the embryo inside its Isolette incubator (pictured in Fig. 4.2).

the maternal calls of their own species over recordings of their own vocalizations. However, although there was no support for the stimulus generalization hypothesis, there was some support for the notion that the birds' exposure to their own vocalizations contributed to identification of their maternal call. Birds given extra prior exposure to their own and broodmate vocalizations via tape recordings before the choice test gave an enhanced response to the maternal call: More of the birds responded in the choice test, they responded more quickly to the maternal call, and they stayed in the region of the maternal call for a longer duration than birds not given the extra exposure. I concluded:

> Though previous auditory stimulation from the neonate's own call enhances the functioning of the auditory perceptual mechanism for species identification in ducklings, it is not yet known whether exposure to the perinatal . . . call is essential to the establishment (acquisition) of the auditory perceptual mechanism. In the latter case, it may prove fruitful to be alert to the possibility that there are stimulative processes which play an important role in establishing perceptual preferences, but may not mimic known forms of learning or conditioning, a possibility previously raised by Gottlieb and Kuo (1965) in relation to a study of prenatal behavioral development in the duck embryo. (Gottlieb, 1966, p. 289)

Thus, preliminaries to an explicit experimental demonstration of probabilistic epigenesis provided these bits of information:

1. Some empirical evidence that the ducklings' auditory experience of their own vocalizations contributes in some way to an identification of the maternal call of the species.
2. That the experiential contribution might not take the form of frank learning (as in stimulus generalization).
3. That the experience may not only enhance responsiveness but might be involved in the establishment of the auditory preference.

FURTHER PRELIMINARIES TO AN EXPLICIT EXPERIMENTAL DEMONSTRATION OF PROBABILISTIC EPIGENESIS: PERFECTING AN EMBRYONIC DEVOCALIZATION PROCEDURE

If I did not have the unusually good fortune of a full-time research position, two full-time research assistants, part-time technical help, and good colleagues, the part of the story I now relate could not have happened.

For 18 months during 1965 and 1966, I struggled to find an innocuous way to mute the mallard duck embryo. I reasoned that if I could mute the embryos and rear them in a soundproof incubator, I could then give them auditory choice tests after hatching to see if their highly specific preference for the mallard maternal call would no longer be in evidence. That was the clearest way to show the importance of hearing their own vocalizations in establishing the preference for the maternal call.

Around 1965, I persuaded Dr. John Vandenbergh to join the basic research group at Dorothea Dix Hospital. John had been trained in zoology and was familiar with making surgical incisions, suturing the wound, and was comfortable with these procedures, whereas, from my experience with Zing-Yang Kuo in 1963, I knew how to open the egg and get at the embryo without doing it harm. While John was waiting for his laboratory to get up and running he agreed to spend some time working with me to see if we could devise an embryonic muting procedure that would allow the bird to hatch and to be otherwise healthy. I had already tried obvious things such as sealing the duckling's bill but that did not prevent it from vocalizing.

As can be seen in Fig. 2.1, the bird's vocal apparatus (syrinx) is in its chest at the base of its trachea. Just beneath the bony syrinx there are the tympaniform membranes, located at the top of each bronchus. The bronchi are tubes that allow air to pass to and from the lungs through the syrinx and trachea, the latter opening into the oral cavity. When the bird forcefully expels air, the internal tympaniform membranes constrict and vibrate, thus causing vocalizations to ensue from the oral cavity. It had been known that the syringeal (tympaniform) membranes were the likely source of sound in birds, but John and I had to rediscover that fact in ducklings in order to figure out how to devocalize the duck embryo. At first we thought that packing wads of sterile cotton around the membranes would prevent them from vibrating and the embryo would be mute. Unfortunately, the only embryos that were muted by this procedure were ones that died because they could not breath. It became clear that we needed to be able to apply some substance to the membranes that would rigidify them but would not hamper the embryo from breathing. My long-time colleague George W. Paulson, a research and clinical neurologist with whom I had collaborated when I was a clinician at Dorothea Dix Hospital, came up with an elegant solution for us. He told us about a nontoxic surgical glue called Collodion that he used in his electroencephalograph work with humans and that was used in the operating room for internal suturing.

Zing-Yang Kuo had left me a large supply of Chinese writing brushes that we had used to apply Vaseline to the opaque egg membranes to render them transparent, thus allowing us to see the embryo for our behavioral studies (Gottlieb & Kuo, 1965). For symbolic reasons, as well as convenience, I used Dr. Kuo's finely tipped brushes to apply Collodion to the duck embryos' tympaniform membranes to render the embryos mute so I

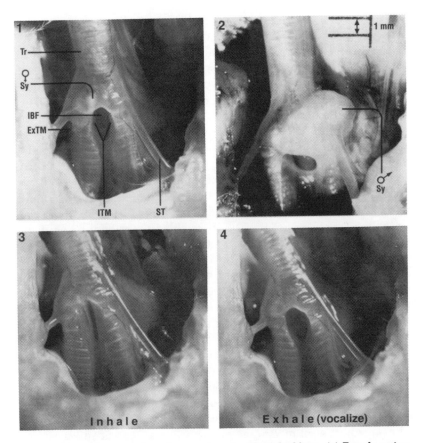

FIG. 2.1. Ventral aspect of vocal apparatus of mallard ducklings. (1) Female syrinx (voicebox). (2) Male syrinx. (3) and (4) Expansion and constriction of internal tympaniform membranes during inhalation and exhalation (with vocalization), respectively. Tr, trachea; Sy, syrinx; IBF, interbronchial foramen; ExTM, external tympaniform membrane (bilateral); ITM, internal tympaniform membranes (bilateral); ST, sterno-trachealis muscle (bilateral).

could study their auditory development in the absence of normal auditory experience. This procedure worked extremely well. Almost all of the embryos regained their voice after a week or so, but I could conduct my tests at 1, 2 and 3 days after hatching with muted ducklings. The embryonic muting procedure was just what my research program required to make a really giant leap forward experimentally in testing the concepts of predetermined and probabilistic epigenesis.

John Vandenbergh and I wanted to share our finding with our scientific colleagues, so I asked him if he wouldn't mind having his name on a paper that not only would describe the embryonic devocalization procedure and thus truly prove experimentally that the syringeal membranes were the source of sound in all species of birds, but one in which I would also describe the range of vocalizations in both duck and chick embryos, which I had been working on for some time in connection with my behavioral research. John agreed, and that work was eventually published in the *Journal of Experimental Zoology* (Gottlieb & Vandenbergh, 1968).

The mallard duck embryo begins to vocalize 2–3 days before hatching, on day 24 of embryonic development, at which time the tip of its bill begins to penetrate into the air space at the large end of the egg and it begins pulmonary (lung) respiration. In perfecting the devocalization procedure, we learned that it was critical for the embryo to be breathing if it was to survive the muting procedure. If we pulled the embryo's bill, head, and neck into the air space prior to day 24, and did nothing further to it, the embryo inevitably expired because it was not yet able to breathe through its lungs. Early on day 24 the embryos can breathe through their lungs so they survive the head-pulling procedure and the devocalization procedure.

Figure 2.2 depicts the various embryonic stages leading to hatching in the duck embryo. We discovered that the embryo begins breathing robustly in the tenting stage, just before its bill penetrates the air space. We resolved to devocalize the embryos during the tenting stage early on day 24. Because the embryos were breathing, they could vocalize, albeit feebly, at this time, so we recorded the number of vocalizations each embryo made during the course of the approximately 10-min devocalization procedure and, as it turned out, this brief exposure to their vocalizations did not influence the test results. The entire procedure involves candling the egg, marking the air space in pencil, removing the shell over the air space, applying sterile water to the innermost shell membrane to render it transparent so we can see the extra-embryonic blood vessels, gently pulling the bill of the embryo through a narrow slit in the shell membrane until its head, neck, and chest are

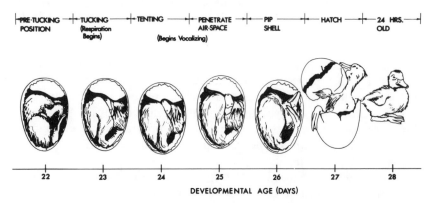

FIG. 2.2.   Late embryonic and early postnatal stages in mallard duckling. (Devocalization takes place in tenting stage.)

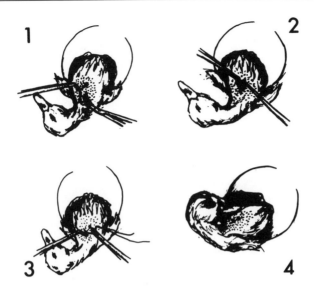

FIG. 2.3. Devocalization procedure. (1) After anesthetic is applied to skin, a small incision is made over the syrinx. (2) Collodion is applied in area of internal tympaniform membranes. (3) and (4) Originally, two sutures closed the wound; later, we simply applied Collodion and held the flaps of skin in place for a brief period while the Collodion hardened.

visible, applying a topical anesthetic to the skin above the syrinx, making a small incision in the skin over the syrinx, visualizing the location of the internal tympaniform membranes, applying Collodion to the membranes, closing the incision, and placing the embryo in its own individual sound-proof incubator. (In later years we made two changes in the procedure: we flooded Collodion into the area below the syrinx via the tip of an injection needle placed in that area, and we closed the incision with Collodion rather than the sutures shown in Fig. 2.3.)

3

Experimental Demonstration of Probabilistic Epigenesis (in Two Parts)

PART ONE

With the embryonic devocalization procedure perfected, the next step was to have individual soundproof incubators constructed so each muted embryo could be deprived of the normal auditory experience of hearing its own or broodmate vocalizations. (In some 10% of the cases, the embryo is not completely devocalized, so it is necessary to have individual soundproof incubators so the completely muted embryos can't hear the vocalizations of incompletely devocalized birds.) The first incubators having been constructed with the excellent help of my friends Clarence Gower, Melvin Humphrey, and David Wilder, the big experiment could now be undertaken: Would the embryonic muting procedure do away with the highly specific auditory perceptual mechanism for species identification?

A number of people helped with the initial experiments, which were labor-intensive because we required a large number of embryos whose health and well-being were carefully monitored from early on day 24 of embryonic development to 1 or 2 days after hatching, at which time all the devocalized birds were tested. Carol Ripley and her husband, Kenneth Ripley, Marieta Barrow Heaton (a graduate student), and Evelyn Hale (née Strickland) assisted with devocalization and monitoring the embryos for signs of vocal recovery. (If an embryo could vocalize at all, it was discarded from the experiment.) Jo Ann Bell (née Winfree) and I tested the muted birds.

The usual time of testing ducklings in other of our laboratory experiments was 24 hr after hatching, so that is the way we started our research with the muted birds. I noticed that as a consequence of having their head pulled out of the egg, the muted birds hatched 10–12 hr earlier than the unopened eggs. This meant they would be developmentally younger than the birds from unopened eggs when both groups were tested at 24 hr after hatching. Further, I thought it possible that the operative procedure, although it seemed minor, might slow the muted embryos' development, so we had to be very careful in weighing these factors in the test results, should an auditory perceptual deficiency be observed in our muted birds.

The first group of muted birds we tested did show a weaker preference for the mallard maternal call versus the wood duck maternal call at 24 hr after hatching, whereas the sham-operated birds, in which a dry brush had been passed over their tympaniform membranes in a "mock" devocalization, showed almost a perfect preference for the mallard call. To get at the question of a developmental lag in the muted birds, we retested these birds at 48 hr after hatching. Indeed, there was a significant improvement in their preference for the mallard call at 48 hr, with almost all of the mute ducklings that had chosen the wood duck call in the first test now preferring the mallard in the second test. This suggested that the relatively poor performance of the mute ducklings in the first test simply reflected a delay in the development of their preference for the mallard call, which was an uninteresting result. Therefore, I decided to give the muted birds their first test at 48 hr after hatching. (These and the results that follow are described in detail in my 1971 monograph, *Development of Species Identification in Birds* [Gottlieb, 1971a].)

When we tested the muted mallard ducklings at 48 hr after hatching with the mallard call versus a duckling call, the mallard call versus a pintail maternal call, and the mallard call versus a chicken maternal call, the muted ducklings were virtually flawless in selecting the mallard over the duckling call and the mallard over the pintail call, but they absolutely flunked in the mallard versus the chicken call test, with 12 showing a preference for the mallard and 9 showing a preference for the chicken. In view of the improvement in the preference for the mallard call shown by the muted ducklings in the mallard versus wood duck test between 24 and 48 hr after hatching, I decided to retest the birds in the mallard–chicken test at 60 hr to see if the departure from their normal preference persisted or would be rectified. In contrast to the earlier result in the mallard–wood duck test, the mute ducklings in the mallard–chicken test continued to show a lack of preference for the mallard. Most importantly, the muted birds were treating the two maternal calls as if they were equivalent; ducklings that preferred the mallard call in the first test were as likely to choose the chicken call as the mallard call in the second test (Gottlieb, 1971a, chapter 10).

I realized that these results, which challenged the basic premise of instinct theory, would be strongly resisted in many quarters unless I could show that the lack of the species-specific preference for the mallard call could not be ascribed to the surgical trauma of devocalization. The routine scientific way to evaluate the effect of the surgery per se is to do a so-called sham operation, which in this case meant opening the egg, pulling the embryo's head out of the shell, applying the anesthetic to the skin, making the incision over the syrinx, drawing a dry brush over the embryo's tympaniform membranes, and so on.

Two of these groups were done: one reared in the soundproof incubators (vocal-isolated), the other reared communally as usual (vocal-communal). Both of these groups showed the usual preference for the mallard call over the chicken call at 48 hr after hatching. The best control for the surgery, however, proved to be a group in which we had accidentally miscalculated the age at devocalization: They were 15–23 hr more advanced and thus had heard their own vocalizations for that period of time before they were muted. When these muted birds were tested at 48 hr after hatching, they showed the usual strong preference for the mallard call over the chicken call (17 chose the mallard call and only 1 chose the chicken call; see Table 3.1).

Another aspect to this last experiment that I thought was exceedingly important was that it showed that merely embryonic exposure to their own "contact call" (the call produced most often by embryos) was sufficient for completely normal auditory perceptual development after hatching.

The reader will have noticed that the muted mallard ducklings "flunked" only the mallard–chicken test and passed all the other tests: mallard–duckling call, mallard–wood duck call, mallard–pintail call.[1] When the

---

[1]I realize that some readers would conclude that the preference for the mallard maternal call shown in those tests is explained by instincts, by which they would mean that auditory experience plays no role in the development of the devocal birds' attraction to the mallard call, at least in these choice tests. Later, I describe the evidence for the contribution of spontaneous (internally generated) neural activity in the development of auditory functioning. In addition to that consideration, for a number of years I tried to discover if nonauditory embryonic experiences contributed to making features of the mallard maternal call attractive. With respect to the preferred repetition rate of the mallard call, for example, I thought that the embryo experiencing its own heart rate at around 3–5 beats per second (around 240 beats per minute) might be a contributing factor. Thus, I incubated eggs at low temperature, thinking that would lower heart rate, but when we examined those embryos' heart rate it was close to normal, even though they hatched on day 30 instead of day 26–27! When we tested these birds after hatching, they preferred the 4-note/sec mallard call over an artificially slowed 2-note/sec mallard call as do normally incubated birds, but, for some unknown reason, they preferred a low-frequency mallard call (one with the higher frequencies filtered out) over the normal mallard call. We replicated this experiment twice with the assistance of Laura Hyatt and Carmen del Real, and got the same results each time. The experiential link here was so peculiar, I did not try to publish the results. But do notice that a change in incubation temperature resulted in a non-species-typical *behavioral* outcome. Thus, I think it will further knowledge of developmental mechanisms at all levels of analysis if we continue to take a probabilistic epigenetic view about development and do not use instinct as an explanatory concept, as many other authors have also suggested.

TABLE 3.1
Preference of Vocal-Communal and Devocal-Isolated Ducklings in Simultaneous
Auditory Choice Tests at 48 hr After Hatching

	Preference	
Prior Auditory Experience	Mallard Call	Chicken Call
Normal vocal, self and sib auditory experience	24*	0
Devocalized, no auditory experience	26	15
Late devocalization, 15–23 hr of embryonic exposure to self-vocalizations	17*	1

*p < .0001, binomial test. (Data from Gottlieb, 1971a, 1978.)

repetition rate and fundamental frequency of the various calls are scrutinized (Gottlieb, 1971a, p. 143), it becomes apparent why the muted ducklings can readily prefer the mallard call over the duckling call and the pintail call: The fundamental frequencies of these calls are radically different from the mallard call. Although the wood duck call shares a fundamental frequency range with the mallard call, the repetition rate of the two calls is very different, thereby providing a basis for the preferences of the mallard. (Indeed, when, in an experiment with vocal ducklings, they were faced with the wood duck call at the same rate as the mallard call, the ducklings showed no preference between the calls.) In the case of the chicken call, both the rate and the fundamental frequency are more similar to the mallard than are any of the other calls. From this way of analyzing the discriminative features of the calls, the duckling and pintail calls can be discriminated from the mallard call on the basis of frequency, whereas the wood duck call can be discriminated from the mallard call on the basis of rate. The chicken call is relatively difficult for the muted ducklings to discriminate from the mallard call because of the similarity in both frequency and rate.

Given that the muted mallard ducklings' usual auditory perceptual skills were present after they heard their embryonic vocalizations for 15–23 hr and absent when they were deprived of that experience, I concluded:

The present results indicate that the epigenesis of species-specific auditory perception is a probabilistic phenomenon, the threshold, timing, and ultimate perfection of such perception being regulated jointly by organismic and sensory stimulative factors. In the normal course of development, the manifest changes and improvements in species-specific perception do not represent merely the unfolding of a fixed or predetermined organic substrate independent of normally occurring sensory stimulation. With respect to the evolution of species-specific perception, natural selection would seem to have involved a selection for the entire developmental manifold, including both the organic and normally occurring stimulative features of ontogeny. (Gottlieb, 1971a, p. 156)

PART TWO

In September of 1971, before the start of the International Ethological Congress in Edinburgh, Scotland, where I planned to announce the results of devocalizing mallard duck embryos, a conference on imprinting was held at the University of Durham, England (Fig. 3.1). The major premise of the concept of imprinting is that young birds cannot identify members of their own species and that such identification is acquired based on their early postnatal experience of following their hen from the nest. Thus, I took a decidedly unpopular course when I reminded my colleagues that, based on my experiments with chicks and ducklings hatched in incubators, the hatchlings were capable of identifying the maternal assembly call of their respective species in advance of contact with the hen. Thus, learning the visual characteristics of their species through contact with the hen was a subsidiary process that took place in the context of the hatchlings' attraction to the maternal call.

Konrad Lorenz had been invited to be a discussant of the papers, and when he came to my paper he blithely reminded all concerned that, during development, there were two sources of information, one innate or instinctive flowing from the genes, the other flowing from the environment (imprinting in this case), and that I had demonstrated the importance of the former. Lorenz had not only deftly deflected my challenge to the significance of imprinting, he had used the opportunity to say how my findings supported the nature–nurture dichotomy as well! There was some well-deserved hilarity at my expense, but I (mistakenly) thought I might redeem myself at the Ethological Congress. Burdened by such a remarkable lack of foresight, it says something for science (and scientists) that I have been able to make any sort of contribution to the field.

When I announced my devocalization findings at the 1971 International Congress of Ethology in Edinburgh, Scotland, even some of my closest associates were grouchy with me. Numerous pints of ale were consumed while Masakazu ("Mark") Konishi and Patrick Bateson in particular tried to

FIG. 3.1. Participants in the Imprinting Conference held at the University of Durham, England, in 1971. (Top from left) H. S. Hoffman (United States), M. W. Schein (United States), E. A. Salzen (Canada), F. Schutz (West Germany), P. Bateson (England). (Middle) K. Immelmann (West Germany), Gilbert Gottlieb (United States), N. Bischof (Switzerland), S. J. Dimond (England), G. MacDonald (Canada), H. B. Graves (United States), W. Sluckin (England). (Bottom) E. Fabricius (Sweden), F. V. Smith (host and convenor, England), E. H. Hess (United States), K. Z. Lorenz (West Germany), G. J. Fischer (United States), L. J. Shapiro (Canada).

educate me in what was required to show that my contention about the role of prenatal experience in the development of instinctive behavior was correct. I was a very recalcitrant student and I am everlastingly grateful to Professors Bateson and Konishi for persevering with me, whether slightly inebriated or sober. They inspired me to begin a course of experiments that ultimately took more than 10 years to resolve in a completely satisfactory manner. I like to think that there was some payback to my two colleagues for taking part in these discussions. Professor Bateson (1976) eventually wrote an article that reflected the central issue that we were all groping for in those discussions: Was the experiential input requirement for instinctive behavior rather general or was it highly specific? For the case at hand, the question was whether a fairly broad range of auditory stimuli would do or whether the input required was the embryo's vocalization.

Eventually, it became clear that in order to prove my hypothesis I must demonstrate a specific acoustic–experiential link between the properties of the embryonic contact call and the critical or distinctive features of the

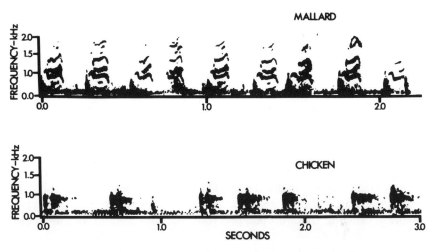

FIG. 3.2. Sonagrams of mallard and chicken maternal calls.

mallard maternal call. I decided to proceed, first, by determining the critical acoustic features that make the mallard call attractive to mallard ducklings and then, second, to determine whether the embryo's experience of particular features of the embryonic contact call was crucial to the duckling's perception of the critical features of the mallard maternal call. The first task proved a lot easier and more straightforward than the second task, because, although there was to be an indubitable link between the embryo's experience of the contact call and its later perception of the distinctive features of the mallard call, the relationship was not linear (obvious) and, therefore, not predictable or readily apparent. Because it is the latter type of discovery that broadens the conceptual scope of one's discipline, I was once again exceedingly fortunate to have the long-term personnel support from the North Carolina Department of Mental Health to allow the full-time pursuit of a goal, the path to which only became clear as unexpected and unwanted experimental results forced me to rethink the problem.

As can be seen in Fig. 3.2, there are two obvious acoustic similarities between the mallard and chicken calls that could be the source of the muted ducklings' "confusion." The calls share a low-frequency component below 1000 Hz, and their repetition rates are fairly close to each other (2.3 notes/sec in the chicken call and 3.7 notes/sec in the mallard call).[2]

---

[2]Acoustic analyses of various mallard and junglefowl (progenitor of domestic chickens) maternal calls recorded in nature indicated the species-typicality of the frequency and repetition rates of the particular mallard and chicken calls used in the laboratory experiments described here (Miller & Gottlieb, 1978, and unpublished observations).

Determining the critical acoustic features of the mallard maternal call was given a substantial assist by Marieta Barrow Heaton, who had already taken on that problem as the topic of her doctoral dissertation. Using vocal mallard ducklings and embryos, and determining their responsiveness to acoustically modified mallard maternal calls, Marieta showed that repetition rate and high-frequency components of the call were the critical acoustic features that made the mallard maternal call attractive to intact mallard ducklings (Heaton, 1971). It remained for me to do two further experimental analyses that would show that (1) the muted birds were using these two dimensions of the mallard call but in a less refined way than the vocal birds, such that they would find the acoustic features of the chicken call as attractive as the mallard call. If that were the case, I would then have to go on (2) to ask the question if exposure to the embryonic contact call is unique in bringing about normal perceptual development in the muted bird.

The second question represents the essence of the probabilistic conception of epigenesis as applied to the particular experimental finding at hand. That is, does natural selection select for the entire developmental manifold (normally occurring experience as well as the organic substrate) or just the genetically predetermined organic (auditory) substrate? If the latter were true, then a broad range of auditory inputs would suffice to keep the endogenously developing auditory mechanism intact. I believe this was the point that Professors Bateson and Konishi were urging on me while we were consuming our ale in Edinburgh. Certainly, Roger Sperry (1971, p. 32) had gone even further than that when he wrote: "In general outline at least, one could now see how it could be entirely possible for behavioral nerve circuits of extreme intricacy and precision to be inherited and organized prefunctionally solely by the mechanisms of embryonic growth and differentiation." Professor Sperry's thinking and research were so highly regarded he was later awarded a Nobel Prize in physiology. Thus, it is no wonder that my zoologically trained colleagues were dubious about my claims of the importance of highly specific, normally occurring prenatal experience in what otherwise appeared to be the instinctive neural and behavioral development of ducklings. The way Sperry and others viewed this kind of apparently innate or instinctive development, any form of auditory stimulation ought to keep the auditory mechanism intact.

If the reader will kindly refer back to Fig. 3.2, it can be seen that the chicken and mallard calls share low-frequency components and relatively slow repetition rates. They differ on other acoustic features such as frequency modulation and amplitude modulation. Because there are these other differences between the calls, to unambiguously rule in degradation in the muted birds' perception of the frequency components and repetition rate of the mallard call, the clearest procedure was to edit these features of the mallard call itself to make it approximate those features in the chicken call and make specific predictions on the preferences of the muted birds in choice tests between the two mallard calls. First I experimentally manipulated the frequency components, and then I manipulated the repetition rate of the mallard call.

As can be seen in Fig. 3.3, the muted birds are deprived of hearing those frequencies of the embryonic contact call between 1500 and 2500 Hz, and these correspond to the upper frequencies of the mallard call. Because the chicken and mallard call share frequencies at 1000 Hz and below, I hypothesized that, in the absence of hearing their own vocalizations, the muted birds were insensitive to the higher frequency components between 1500 and 2500 Hz in the mallard call and, thus, responded to the mallard and chicken calls as if they were the same call (i.e., they find the two calls equally attractive).

As shown in Fig. 3.4, by filtering out the frequencies above 825 Hz in the mallard call, its frequency composition is similar to the chicken call. When we tested the muted birds for their preference between the filtered and unfiltered mallard calls, in conformity with the hypothesis they were indeed less sensitive than the vocal birds to the absence of the high frequencies (Gottlieb, 1975a). To see if we could prevent this perceptual deficit, in the next experiment (Gottlieb, 1975b), we exposed another group of muted birds to a tape recording of the embryonic contact call for 5 min/hr from embryonic day 24, 1800 hr, through embryonic day 26, 1800 hr, of

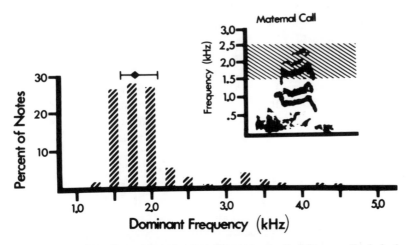

FIG. 3.3. Dominant frequency of 444 call notes recorded from mallard duck embryos on days 25 and 26 of development. Because embryos emit few alarm-distress calls, and those calls have a dominant frequency (most of their energy) above 3.0 kHz, approximately 90% of the 444 notes in this sample are contact notes (i.e., notes whose dominant frequency is between 1.5 and 2.5 kHz). From Scoville, 1982, Fig. 2.9, p. 69. Reprinted by permission.

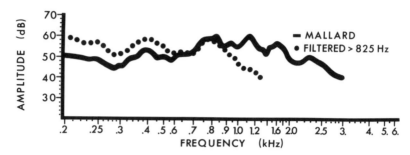

FIG. 3.4.   Frequency analysis of normal mallard maternal call and same call with frequencies above 825 Hz filtered out.

development. The stimulated mute birds not only showed a stronger preference than the unstimulated mute birds for the normal call, it was the same degree of overwhelming preference as the vocal-communal birds. To test the experiential specificity hypothesis, we stimulated yet another group of mute birds with low-frequency white noise between 300 and 1800 Hz, and their preference was not different from the unexposed mute birds. In another test, we stimulated the muted birds with their alarm-distress call, which has its main energy above 3000 Hz, and it had no effect (they performed as poorly as the unstimulated mute birds).

In summary, only prenatal stimulation with the embryonic contact call prevented the muted ducklings' high-frequency perceptual deficit after hatching, which is consistent with the probabilistic view of epigenesis. The acoustic characteristics of the contact call, white noise, and alarm-distress calls are shown in Fig. 3.5.

Things were looking quite good at this point, but we were not quite home. The birds just mentioned were tested at 24 hr after hatching, and when we retested them at 48 hr after hatching they gave us equivocal results. Because the original finding of a perceptual deficit in the muted birds in the chicken–mallard call test was at 48 hr after hatching, the equivocal retest results required that we repeat the experiments with initial testing at 48 hr rather than 24 hr after hatching.

When we repeated the experiment at 48 hr (Gottlieb, 1975c), much to our surprise the unstimulated muted birds performed as well as the vocal birds in preferring the normal mallard call over the filtered mallard call. Interestingly, when we tested yet another group of mute birds at 65 hr after hatching, they showed the same inferior performance as the mute birds tested at 24 hr (Fig. 3.6). When we stimulated devocalized embryos with the contact call, the deficits at both ages were corrected. As Fig. 3.6 indicates, the devocal birds stimulated with the embryonic contact call performed as well as the vocal birds at 24 and 65 hr. So, embryonic auditory experience

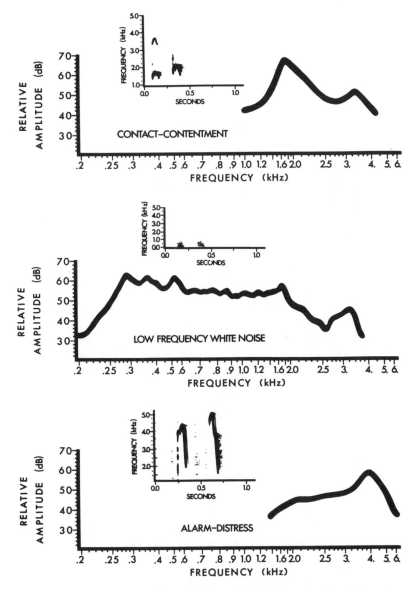

FIG. 3.5. Acoustic analyses of contact call, low-frequency white noise, and alarm-distress call.

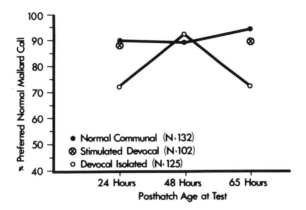

FIG. 3.6. Proportion of vocal-communal and devocal-isolated ducklings that preferred the normal maternal call in the normal mallard maternal call versus >825 Hz attenuated mallard maternal call test at 24 hr, 48 hr, and 65 hr after hatching. Note that the devocal-isolated ducklings that were stimulated with a recording of the embryonic contact call performed as well as the normal-communal ducklings at 24 hr and 65 hr after hatching.

cannot only prevent the deficit at 24 hr, but it also prevents the deterioration at 65 hr after hatching.

Because the original deficit with the chicken call occurred at 48 hr, the present results, although important in their own terms, do not explain the mute birds' deficit in the mallard–chicken test at 48 hr. We would have to explore the mute birds' perception of repetition rate to see if that explained the original test results.

To test repetition rate, as shown in Fig. 3.7, we simply enlarged the spaces between the notes of the normal mallard call so it pulsed at the same rate as the chicken call (2.3 notes/sec), and tested muted birds for their preference for the normal mallard (3.7 notes/sec) versus the slowed mallard (2.3 notes/sec) at 48 hr after hatching (Gottlieb, 1978). Bingo!

As the summary in Table 3.2 shows, the preference figures for the mallard–chicken tests for the vocal and muted birds are similar to those in the normal mallard–slowed mallard choice tests: The vocal birds show an overwhelming preference for the mallard over the chicken, whereas the mute birds do not show a preference; in the normal–slowed mallard test, the vocal birds show an overwhelming preference for the normal mallard, whereas the mute birds do not show a preference.

We were now 7 years down the road since the inspiring discussions with Professors Bateson and Konishi in Edinburgh. I must describe yet two more twists before getting things really right.

I next did the obvious experiment of giving devocalized embryos prenatal exposure to the embryonic contact call at 4 notes/sec and testing them for

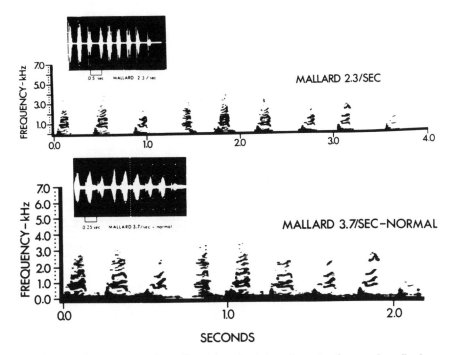

FIG. 3.7. Sonagrams and oscillographic depiction (insets) of normal mallard maternal call (3.7 notes/sec) and slowed mallard maternal call (2.3 notes/sec). (Notes in slowed mallard call appear shorter than in normal call because of the necessity of greater photographic reduction.)

their preference in the normal mallard versus slowed mallard calls at 48 hr after hatching (Gottlieb, 1980a). In order to determine how specific the embryonic stimulation had to be, we also exposed other groups of devocalized embryos to the contact call at either 2.1 notes/sec or 5.8 notes/sec. There were several possible outcomes: (a) It could be that exposure to any of the contact calls would foster a preference for the normal mallard call; (b) it could be that only exposure to the 4-notes/sec contact call would bring about the normal preference; (c) exposure to the 2-notes/sec contact call might induce a preference for the slowed maternal call over the normal maternal call. As before, the devocalized embryos were exposed to one of the contact calls for 5 min/hr from embryonic day 24, 1800 hr, through embryonic day 26, 1800 hr, and they were tested at 48 hr after hatching. The embryonic contact calls are shown in Fig. 3.8.

As shown in Table 3.3, the only devocal group to show a preference for the normal mallard call was the one exposed to the contact call at 4 notes/sec; exposure to the contact call at the other rates was ineffective in promoting the normal preference, and exposure to the 2-notes/sec contact

TABLE 3.2
Preference of Vocal-Communal and Devocal-Isolated Ducklings in Simultaneous
Auditory Choice Tests at 48 hr After Hatching

Group	Preference	
	Normal Mallard Call (3.7 notes/sec)	Normal Chicken Call (2.3 notes/sec)
Vocal-communal	24*	0
Devocal-isolated	26	15

Group	Normal Mallard Call (3.7 notes/sec)	Slowed Mallard Call (2.3 notes/sec)
Vocal-communal	30*	2
Devocal-isolated	20	16

*$p < .0001$, binomial test. (Data from Gottlieb, 1978.)

TABLE 3.3
Preference of Vocal-Communal and Devocal-Isolated Ducklings in
Simultaneous Auditory Choice Test With Mallard Maternal Call at Normal and
Slowed Repetition Rates After Embryonic Exposure to 2.1-Notes/Sec, 4-Notes/Sec, or
5.8-Notes/Sec Embryonic Call

Auditory Experience	Preference	
	Normal Mallard Call (3.7 notes/sec)	Slowed Mallard Call (2.3 notes/sec)
Vocal-communal		
Normal control group	30*	2
Devocal-isolated		
No vocal exposure	20	16
Exposed to 2.1-notes/sec embryonic call	14	14
Exposed to 4-notes/sec embryonic call	28*	5
Exposed to 5.8-notes/sec embryonic call	16	16

*$p < .0001$, binomial test. (Data from Gottlieb, 1980a.)

call did not result in a preference for the slowed maternal call. The latter result suggests a lack of plasticity, which, as it later turned out, is misleading—I shall describe just how malleable the embryo and hatchling can be under different rearing conditions in chapter 7.

In the design of this last experiment, I was still thinking in a linear way, and the results seemed to support the view that exposure to a 4-notes/sec contact call would be required to foster the preference for the normal mallard call (3.7 notes/sec). Vocal embryos do produce 4-notes/sec contact calls, but, as we learned later, the embryo's production of the contact call is highly variable before hatching, and includes 2 notes/sec up to 6 notes/sec (Scoville, 1982). What contributed to my linear thinking was the

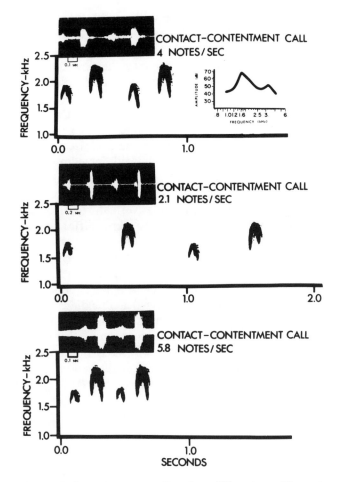

FIG. 3.8. Embryonic contact call at three different repetition rates.

mistaken impression that the embryo's responsiveness to the mallard maternal call was highly specific in advance of auditory experience. We had tested embryos on day 22 of development (2 days before embryonic vocalization begins) and found that they gave a behavioral response to the mallard maternal call only when it was pulsed at 4 notes/sec, and not when it was pulsed at 1, 2, or 6 notes/sec (Gottlieb, 1979). However, when we subsequently examined the embryos' responsiveness to other rates between 2 and 6 notes/sec, it turned out that 2 and 6 notes/sec were the only rates to which they were unresponsive! On day 22, the aurally inexperienced embryos were responsive to 2.8, 3.7, 4.7, 5.1, and 5.5 notes/sec. (These previously unpublished results are presented in chapter 4, along with a discussion of the precise role of experience in the present case.)

I learned of the incorrectness of my linear thinking when, during the course of a "critical period" experiment, I found (Gottlieb, 1981) that the devocal embryos stimulated with the 4-notes/sec contact call did not show a preference for the normal mallard call over the slowed mallard call at 24 hr after hatching, but only at 48 hr after hatching (results in Table 3.4). Because vocal birds prefer the normal mallard at 24 hr as well as 48 hr after hatching, something was not right. Once again, I took the core proposition of probabilistic epigenesis to heart as it applies to the present problem: "In the evolution of species-specific perception, natural selection has involved a selection for the entire developmental manifold, including both the organic and normal occurring stimulative features of ontogeny" (Gottlieb, 1971a, p. 156; 1975b). Based on Scoville's (1982) observations of the vocalizations of embryos, the normally occurring stimulative features of ontogeny for the mallard duck embryo would be exposure to all the different repetition rates of the embryonic contact call, not just 4 notes/sec.

So, I repeated (Gottlieb, 1982) the previous experiment with devocalized embryos, this time giving each one of them exposure to 2.1, 4, and 5.8 notes/sec of the contact call, as happens under normal conditions of development. Lo and behold, now the devocal ducklings showed the usual preference for the normal mallard call over the slowed mallard call at 24 hr after hatching (Table 3.5). To further pursue the question of the specificity of the experiential input requirement, we asked if the embryos had to merely experience the rhythmic component of the contact call by exposing them to a narrow band of unmodulated white noise (1500–2500 Hz) pulsed at 4 notes/sec and compared their responsiveness at 48 hr after hatching with the devocal birds that had been exposed to the frequency-modulated 4-notes/sec contact call. (The calls are shown in Fig. 3.9.)

TABLE 3.4
Preference of Devocal-Isolated Ducklings in Simultaneous Auditory Choice Test With Mallard Maternal Call at Normal and Slowed Repetition Rates After Embryonic Exposure to 4-Notes/Sec Embryonic Contact Call (5 min/hr Exposure)

	Preference	
Auditory Experience	Normal Mallard Call (3.7 notes/sec)	Slowed Mallard Call (2.3 notes/sec)
Embryonic exposure from day 25, 0800 hr, to day 26, 0800 hr; test at 24 hr after hatching[a]	21	12
Same, except test at 48 hr after hatching	22[*]	9

[*]$p = .03$, binomial test. (Data from Gottlieb, 1981.)
[a]Because this result was in the direction of statistical significance, we subsequently added more birds to this condition to validate the interpretation of the ineffectiveness of the experience; those results supported the interpretation and are shown in Table 3.5 (27 vs. 20).

TABLE 3.5
Preference of Vocal-Communal and Devocal-Isolated Ducklings in Simultaneous
Auditory Choice Test With Mallard Maternal Call at Normal and Slowed Repetition
Rates at 24 hr After Hatching (5 min/hr Embryonic Exposure to Contact Call From
Day 25, 0800 hr, to Day 26, 0800 hr)

	Preference	
Auditory Experience	*Normal Mallard Call* *(3.7 notes/sec)*	*Slowed Mallard Call* *(2.3 notes/sec)*
Vocal-communal		
Normal control	26**	3
Devocal-isolated		
No exposure to calls	26	14
Embryonic exposure to 4-notes/sec		
contact call	27	20
Embryonic exposure to 2.1-, 4-,		
and 5.8-notes/sec contact calls	21*	8

* = .02, **p < .00006, binomial test. (Data from Gottlieb, 1982.)

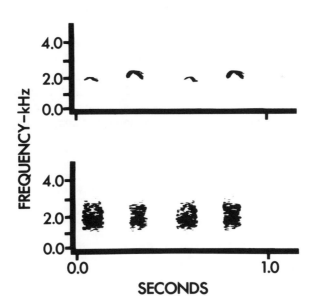

FIG. 3.9. Embryonic contact call (top) and narrow-band white noise pulsed at
same repetition rate (4 notes/sec).

As can be seen in Table 3.6, the narrow-band white noise was ineffective, whereas exposure to the contact call was effective in fostering the preference for the normal mallard call. Thus, the specificity of the experiential input requirement would appear to be total: The devocal embryos cannot be merely exposed to the rhythmic component of the contact call, but must also experience the characteristic frequency modulation of the call if their behavior after hatching is to be normal.

To finish this rather long intellectual journey that began in a bar in Edinburgh in 1971, I decided to do a rather routine experiment to see if there was a delimited "critical period" during which the embryo must experience the contact call, or whether the beneficial effects of exposure to the contact call could occur after hatching as well. Another finding that came out of Richard Scoville's (1982) doctoral dissertation is pertinent here, namely, that the contact call not only becomes much less variable in repetition rate after hatching, but its frequency range also changes so that the call becomes higher pitched, putting the large majority of the notes out of the range of the upper frequency bands (1500–2500 Hz) of the mallard maternal call (see Fig. 3.10). Thus, during the usual course of development, the embryo would be exposed to its highly variable contact call in the frequency range between 1500 and 2500 Hz, whereas after hatching the contact call becomes less variable in repetition rate and higher pitched.

What we did in this experiment (Gottlieb, 1985) was to expose mute embryos and hatchlings to the variable contact call (I'll call it 2, 4, 6 notes/sec from now on) and test them 24 hours after the exposure was terminated. To keep the intervals the same in each group, this means that the embryos were tested at 24 hr after hatching, and the hatchlings were

TABLE 3.6
Preference of Devocal-Isolated Ducklings in Simultaneous Auditory Choice Test With Mallard Maternal Call at Normal and Slowed Repetition Rates After Exposure to 4-Notes/Sec Contact Call or 4-Notes/Sec White Noise

| | Preference | |
| | Normal Mallard Call | Slowed Mallard Call |
Auditory Experience	(3.7 notes/sec)	(2.3 notes/sec)
Vocal-communal		
Normal control	30**	2
Devocal-isolated		
No exposure to calls	20	16
Embryonic exposure to		
4-notes/sec contact call	22*	9
Embryonic exposure to		
4-notes/sec white noise	19	15

*$p = .03$, **$p < .00006$, binomial test. (Data from Gottlieb, 1982.)

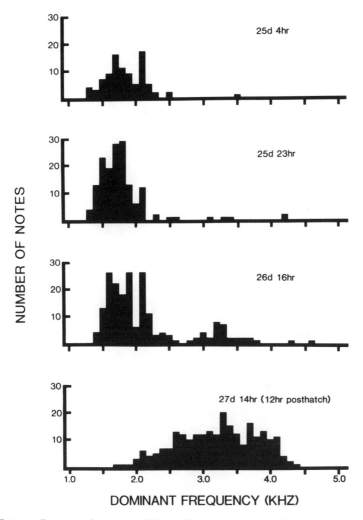

FIG. 3.10. Dominant frequency of the embryos' vocalizations (day 25, 0400 hr, day 25, 2300 hr, day 26, 1600 hr) and the hatchlings' vocalizations (day 27, 1400 hr). There is a remarkable shift in dominant frequency from the embryonic to the early postnatal period. From Scoville (1982). Reprinted by permission.

tested at 48 hr after hatching. As shown in Table 3.7, the ducklings stimulated as embryos chose the normal mallard call over the slowed mallard call, but the ducklings stimulated as hatchlings did not show a preference between the two maternal calls (i.e., the experience was ineffective in the hatchlings). Because the repetition rate of the contact call produced by hatchlings is in the range 4–6 notes/sec (not 2, 4, 6), we stimulated another group of mute hatchlings with the contact call at 4, 6

TABLE 3.7
Preference of Vocal-Communal and Devocal-Isolated Ducklings in Simultaneous
Auditory Choice Test With Mallard Maternal Call at Normal and Slowed Repetition
Rates: Embryonic Critical Period

		Preference	
Auditory Experience	Age at Test	Normal Mallard (3.7 notes/sec)	Slowed Mallard (2.3 notes/sec)
Vocal-communal, normal control	24hr	26**	3
	48hr	30**	2
Devocal-isolated, no exposure	24hr	26	14
	48hr	20	16
Devocal-isolated, embryonic exposure to variable-rate contact call from day 25, 0800 hr, to day 26, 0800 hr	24hr	21*	8
Devocal-isolated, postnatal exposure to variable-rate contact call from day 26, 0800 hr, to day 27, 0800 hr	48hr	14	20
Devocal-isolated, postnatal exposure to less variable (4, 6 notes/sec) contact call from day 26, 0800 hr, to day 27, 0800 hr	48hr	20	15
Devocal-isolated, postnatal exposure to variable-rate contact call from day 26, 0800 hr, to day 27, 0800 hr	24hr	16	11
Devocal-isolated, embryonic exposure to less variable (4, 6 notes/sec) contact call from day 25, 0800 hr, to day 26, 0800 hr	24hr	16	15
Devocal-isolated, embryonic exposure to less variable (2, 4 notes/sec) contact call from day 25, 0800 hr, to day 26, 0800 hr	24hr	17	15

*$p < .02$, **$p < .00006$, binomial test. (Data from Gottlieb, 1985.)

notes/sec prior to testing, and that also was ineffective (line 7 of results shown in Table 3.7). To make sure it wasn't the time interval between exposure and testing that was causing the lack of effectiveness of the variable (2, 4, 6 notes) contact call, we stimulated yet another group of hatchlings as before and tested them at 24 hr instead of 48 hr after hatching. Once again, the exposure was ineffective (line 8 in Table 3.7). Finally, to determine the importance of being exposed to the entire range of repetition rates of the contact call, we exposed embryos to either 2, 4 notes/sec or 4, 6 notes/sec of the contact call and tested them at 24 hr, as we did in the first experiment. Although exposure of the embryos to 2, 4, 6 notes/sec was effective, exposure to the less variable contact calls was ineffective (the bottom two lines of results in Table 3.7).

So, it seems fair to conclude that (a) not only are the embryo's auditory experiential requirements met solely by the acoustic variations characteristic of the embryonic period, but (b) their sensitivity to the experience of these vocalizations is also restricted to the embryonic period if their species-typical preference is to be present at the usual early age (24 hr) after hatching.

4

Perceptual Capabilites of Mallard Duck Embryos: Clarifying the Role of Experience

The foregoing analysis shows the value of a probabilistic view of behavioral epigenesis, particularly with respect to behavior that shows the classical defining features of instinct: species-typical or species-specific, lack of dependence on known forms of learning, adaptive (survival value), and responsive to a narrow configuration of sensory stimulation ("sign stimuli" or "releasers") in the absence of prior exposure to that stimulation (e.g., the maternal assembly call of the species). I would not have had a conceptual basis to motivate the experiments without the developmental framework supplied by a probabilistic view as an alternative to the predetermined epigenesis of behavior.

In order to know the precise role that experience is playing in the development of the mallard hatchling's behavioral attraction to the critical acoustic feature (repetition rate) of the mallard maternal call, it is necessary to examine the embryo's responsiveness to the maternal call before and after it has heard the embryonic contact call. Is the aurally naive embryo already homed in on $4 \pm .5$ notes/sec, as I believed earlier (Gottlieb, 1979), or is its auditory system more broadly tuned to a wider range of rates, and does it become more finely tuned only after it has experienced the contact call?

To answer these questions, we tested embryos on day 22 (aurally naive) and day 26 (aurally experienced) of development with the mallard maternal call pulsed at various repetition rates. The aurally naive embryos were kept in incubators in which there were no advanced embryos, whereas the aurally experienced embryos were kept in incubators in groups of at least 20 vocal embryos. For the latter group, recall that the embryos begin vocalizing on day 24, and they pip the shell on day 25 (which allows them to clearly hear others vocalize as well as themselves), so they would have had considerable normally occurring vocal experience by the time they are tested on day 26.

In the 1960s, with the excellent assistance of Patricia Bush Willis, I had worked out an objective way of recording the behavior of embryos. As can be seen in Fig. 4.1, we affixed tiny needle electrodes to the bill to record bill-clapping and to the body to record heart rate. In the older embryos, we were also able to record their vocal response to the maternal call. The procedure was to record baseline activity for 60 sec prior to exposing them to the call for 60 sec, and then recording their poststimulation activity for 60 sec. For any group of birds so tested, we could determine whether they showed a statistically significant change in activity from baseline during the stimulation period and further, if so, whether their activity returned to baseline in the poststimulation period.

By 1968, it had become clear that, of the three measures, on day 26 of

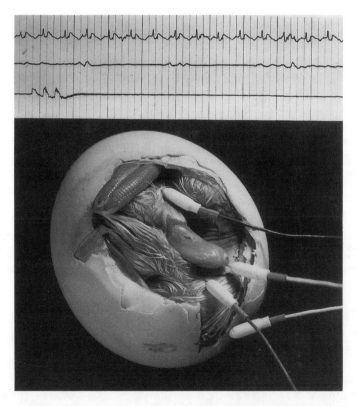

FIG. 4.1. Mallard duck embryo with recording electrodes in place on day 26. Heartbeat is shown on top line, second line depicts bill-clapping, and the embryo's vocalizations are shown on bottom line. Note that bill-clapping and vocalization are independent measures because the embryo can vocalize without making bill movements.

embryonic development, bill-clapping was the most specific, vocalization was somewhat less specific, and heart rate change was nonspecific. What I mean by that is that heart rate would change in response to the playing of the maternal call of any species, vocalization would change in response to only species calls (maternal and peer vocalizations), and bill-clapping would change only to the maternal call of the species. Because the day-22 embryos cannot vocalize, that would be yet another reason to focus on bill-clapping in comparing the responsiveness of day-22 and day-26 embryos to the mallard maternal call at various repetition rates.

To answer the question at hand, we examined the bill-clapping responsiveness of the embryos to the mallard maternal call played at repetition rates of 1/sec, 2.3/sec, 2.8/sec, 3.3/sec, 3.7/sec, 4.2/sec, 4.7/sec, 5.1/sec, 5.5/sec, and 6/sec. Each bird was tested only once with one of the calls. (In nature, when calling their young from the nest, mallard hens utter maternal calls in the range from 3.5/sec to 4.7/sec; Miller & Gottlieb, 1978.)

FIG. 4.2. Embryonic behavioral and physiological test chamber. Carol Ripley observes the embryo and listens for its vocalizations, while the EEG machine gives a graphic record of the embryo's heart rate, bill-clapping, and vocalization on three channels (to the right of Carol's head). The newborn baby Isolette incubator's front and top doors are open for photographic purposes.

TABLE 4.1

Bill-Clapping Responses of Mallard Duck Embryos to Various Repetition Rates of the
Mallard Call

Rates of Maternal Call	Age of Embryos	
	Aurally Deprived, day 22	Aurally Experienced, day 26
1.0/sec	0	–
2.3 /sec	0	–
2.8/sec	–	0
3.3/sec		0
3.7/sec	–	+
4.2/sec		+
4.7/sec	–	+
5.1/sec	–	
5.5/sec	–	
6.0/sec	0	+

Note. +, Statistically reliable increase from baseline; 0, no change; –, statistically reliable decrease from baseline activity. The responses to 1, 2.3, 3.7, and 6/sec are derived from actual quantitative measurements in Gottlieb (1979); the measurements on which the others are based are unpublished.

The embryonic testing setup is shown in Fig. 4.2.

In order to understand the results of this experiment, I must describe how the bill-clapping response itself develops from day 22 to day 26. On day 22 the embryos give only an inhibitory (behavioral freezing) response, whereas by day 26 the aurally experienced embryos give excitatory responses to calls they find attractive after hatching and inhibitory responses to calls they do not approach after hatching. For example, vocal hatchlings freeze to the mallard call at 1 and 2.3 notes/sec after hatching, that repetition rate being in the range of the mallard maternal alarm call (Miller, 1994). Because day-22 embryos are not positively responsive to any call, we must use their freezing response (–) as an indication of their range of responsiveness to the various maternal calls on day 22 of development and compare that to the positive and negative responsiveness of the aurally experienced day-26 embryos.

As can be seen in Table 4.1, although the aurally naive day-22 embryos were responsive (–) to mallard maternal calls in the range from 2.8/sec to 5.5/sec, the day-26 embryos are (+) responsive to a narrower range (3.7–6/sec), which more closely approximates the range uttered by hens in nature (3.5–4.7/sec; Miller & Gottlieb, 1978). Furthermore, it is especially important that the day-26 embryos become negatively responsive (–) to 2.3 notes/sec and positively (+) responsive to 3.7 notes/sec, because (a) they prefer the normal 3.7 mallard call to the 2.3 call after hatching, but (b) they are as positively attracted to the 2.3 call as the 3.7 call after hatching if they have been devocalized (chapter 3). This latter result suggests not only an absence of the narrowing of repetition rate specificity, but also a broadening of specificity from day 22 of embryonic development to 48 hr after hatching in the absence of normally occurring embryonic auditory experience.

Admittedly, the inference that perceptual development is arrested from day 22 to day 26 of embryonic development in the absence of normally occurring auditory experience is not as straightforward as one would like, because the day-22 embryos give only an inhibitory response and the day-26 embryos give both freezing and excitatory responses. To clarify the situation, embryos were placed in auditory isolation on day 23 and tested on day 26 for their bill-clapping response to the mallard maternal call at 1/sec, 2.3/sec, 3.7/sec, and 6/sec.

As shown in Table 4.2, although the partially aurally deprived embryos (they could hear their own voice) do develop an inhibitory response to the maternal call at 1/sec and 2.3/sec (as do the communally incubated embryos), they do not develop the excitatory response to 3.7/sec or 6/sec as do the communally incubated embryos. Because they are vocal and, thus, are subject to auditory self-stimulation, there is some development in their perceptual capabilities in the embryonic period, but it is insufficient to develop the normal positive or excitatory response on day 26. That the partially deprived embryos become negatively responsive to 2.3/sec by day 26 is important for our understanding of the meaning of that response on day 26. The isolated embryos that were allowed to hear themselves continued to be unresponsive to 2.3/sec after hatching (delayed devocalization group in chapter 3), whereas the isolated embryos devocalized at the usual time developed a positive responsiveness to 2.3/sec in the mallard (3.7/sec) versus the chicken (2.3/sec) call test after hatching.

To strengthen the case that mallard embryos' normal perceptual capabilites are substantially stunted if not totally arrested in the absence of normally occurring auditory experience, we have other data from an auditory neurophysiological experiment on the development of vocal-communal and devocal-isolated ducklings' responsiveness to pure tone

TABLE 4.2

Bill-Clapping Responses of Partially Aurally Deprived and Fully Aurally Experienced Mallard Duck Embryos to Various Repetition Rates of the Mallard Call

	Age of Embryos	
Rates of Maternal Call	Partially Aurally Deprived Day 26	Aurally Experienced Day 26
1.0/sec	−	−
2.3/sec	−	−
3.7/sec	0	+
6.0/sec	0	+

Note. Same symbols as Table 4.1. Derived from Gottlieb (1979).

stimulation (Dmitrieva & Gottlieb, 1994). This study examined the influence of auditory experience on the development of brainstem auditory-evoked responses (BAER) to pure tones (a neurophysiological measure of frequency [Hz] sensitivity) in mallard duck embryos and hatchlings.

The experimental setup for the auditory neurophysiological research is shown in Figs. 4.3 and 4.4.

We examined the latency and threshold of the first wave of the BAER in two groups of birds: ones that had been muted on day 24 and kept in auditory isolation (auditory deprivation group) until tested, and vocal embryos that had been exposed to tape recordings of the embryonic contact call (stimulation with CTs in Fig. 4.5). Because the latency and threshold findings were highly similar, I present only the threshold findings.

As can be seen in the top of Fig. 4.5, complete auditory deprivation virtually arrested neurophysiological auditory development beyond day 24 of embryonic development. On day 24, prior to vocal-auditory experience, the embryos are most sensitive to 1000, 1500, and 2000 Hz, and subsequent to day 24, their threshold drops only slightly in the absence of auditory

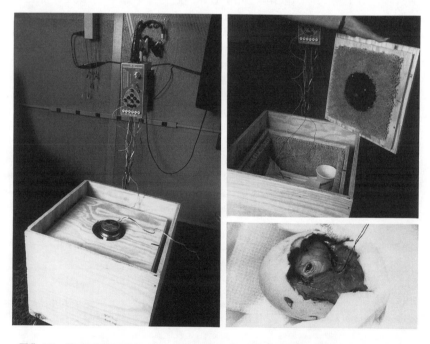

FIG. 4.3. Experimental setup for recording brainstem auditory-evoked responses in embryos and hatchlings (left and right). Day-24 duck embryo (right, below). Setup was in a soundproof chamber adjacent to room that housed a large computer (clinical audiometer), which delivered pure tones and recorded the auditory-evoked responses. (See Fig. 4.4.)

FIG. 4.4. Lubov Dmitrieva seated at clinical audiometer with an auditory-evoked response from a day-24 embryo on the screen.

experience. On the other hand, as can be seen in the bottom of Fig. 4.5, when vocal birds are exposed to the embryonic contact call (CT), their thresholds plunge to their most mature level in the 2 days prior to hatching and show no further improvement in the 2 days after hatching despite continued exposure to tape-recorded CTs, their own and sibling vocalizations. These results fit remarkably well with the behavioral embryonic critical period results described in chapter 3. In addition, they show a virtual arrest of perceptual development in the absence of normal auditory experience, which was also suggested in the behavioral experiments with repetition rate.

DEVELOPMENTAL CONTRIBUTION OF SPONTANEOUS (ENDOGENOUS) ACTIVITY IN THE AUDITORY SYSTEM

Although this next point is a little bit tangential, this seems a good place to deal briefly with the issue of defining experience more broadly to include, in the present case, the known spontaneous (not externally evoked) activity of the avian auditory system. As noted in the previous discussion, there was some, albeit minor, improvement in the hearing thresholds of the aurally deprived embryos and hatchlings described in the preceding section (Fig. 4.5). In chapter 3, I also described behaviorally measured perceptual

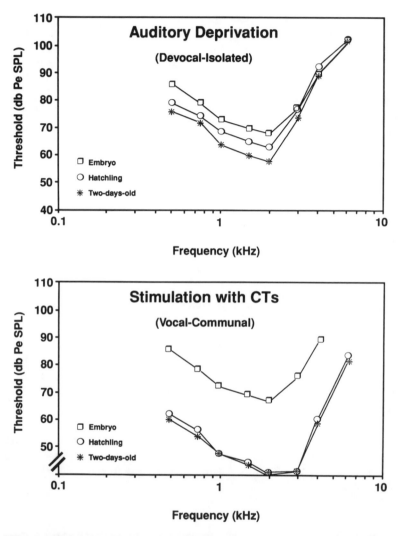

FIG. 4.5. Threshold of brainstem auditory-evoked response in aurally deprived (top) and aurally experienced (bottom) day-24 embryos, hatchlings, and 2-day-old ducklings. Data from Dmitrieva and Gottlieb (1994).

improvement in frequency sensitivity in the devocal-isolated mallard ducklings tested with normal and filtered mallard maternal calls (results in Fig. 3.6).

It is well known that there is nonevoked, spontaneous neuronal electrical activity in all sensory systems studied to date, and the avian auditory system is no exception. It is also known that purely endogenous neural electric activity can further the maturational development of the brain so that spontaneous firing is sometimes recognized as "an epigenetic factor in brain development" (review in Corner, 1994). As that author pointed out, however, although everyone accepts spontaneous firing as a phenomenon, it is an "insufficiently appreciated . . . principle in the development and function of neural networks" (Corner, 1994, p. 3).

The evidence for the phenomenon of spontaneous electrical activity in the avian auditory system comes from the elegant developmental neurophysiological studies of chick embryos and hatchlings by Edwin Rubel and his many collaborators (reviewed in Rubel & Parks, 1988). As shown in Fig. 4.6, the nucleus magnocellularis, one of the cochlear nuclei in the periphery of the avian auditory system, is capable of both spontaneous and evoked activity (left and right sides of Fig. 4.6, respectively). When the tympanic membrane is punctured, the evoked activity persists, but when the columella is removed (necessary part of the receptor apparatus for external auditory stimulation), the evoked activity is no longer present even though the capability for spontaneous firing persists. And when the cochlea is removed the spontaneous activity itself disappears. (The small amount of electrical activity that remains is generated by the electronic recording apparatus used in the neurophysiological experiments—lower right panel in Fig. 4.6.)

Although the constructive role of endogenously produced neural activity in the development of the nervous system is, as Corner observed, "underappreciated," I would nonetheless like to put it forth in the present case as the necessary influence to the further maturation of frequency sensitivity in the devocalized mallard duck embryos and hatchlings described here and in chapter 3. That is the main reason why I have been suggesting since the early 1970s that experience should be more broadly defined to include activity produced within the organism itself (endogenous motor as well as sensory-system activity). I think it is clear that immature nerve cells benefit from activity, whether the source of their excitation is inside or outside the animal. Because neural development is enhanced by both sources, it has always seemed reasonable to me to include spontaneously generated activity as part of the organism's experience.

As a matter of fact, it is only by denying (or not acknowledging) the role of spontaneous endogenous activity within the nervous system as playing a formative role in neural and behavioral development that the outmoded nature versus nurture conception can be kept alive, as witness a recent paper in the *Journal of Neuroscience*. The authors found adult-like organization of the eye dominance columns in the brains of newborn monkeys in advance of visual exposure to light. With reference to their finding, Horton and Hocking (1996, p. 1806) wrote: "The catch phrase 'nature versus nurture' encapsulates a debate [!] central to developmental biology . . . . Evidently,

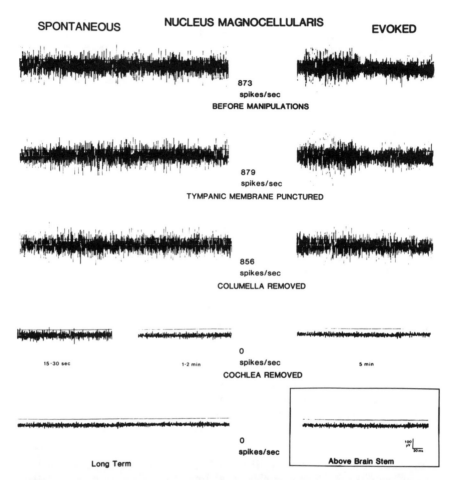

FIG. 4.6. Spontaneous and evoked auditory activity recorded in nucleus magnocellularis, one of the cochlear nuclei in birds. From Rubel and Parks (1988).
Reprinted by permission.

the functional architecture of the macaque visual cortex develops according
to an innate program, independent of light or visual stimulation."

It is certainly important to know that light stimulating the eyes does not
play a necessary role here, but these authors don't even mention the
probable contribution of spontaneous firing in the visual system, which
blunts our understanding of the developmental process. When one writes,
as I have here and elsewhere, that the nature–nurture conception is
outmoded, it represents wishful thinking after all these years of reading and
writing critiques and doing experiments on this very topic. I include
reference to Horton and Hocking's recent publication in the highly re-

spected *Journal of Neuroscience* to indicate that I am not beating a dead horse. Several years ago, Johnston (1987) documented the continuing wide usage of nature or nurture as singular explanatory concepts in developmental studies of animal behavior. In a similar vein, Miller (1988, in press) showed that although the concept of instinctive behavior, as defined here, is a valid and useful one, the absence of frank learning or practice does not mean that experience does not play a role in the ontogeny of instinctive behavior.

ROLE OF EXPERIENCE

With regard to the role of normally occurring auditory experience, the experiments with aurally deprived and aurally experienced embryos allow us to conclude that experiencing their own vocalizations and those of peers enhances the frequency sensitivity and fine tunes the repetition-rate specificity of the mallard duck embryo.

The preceding results and those of others supportive of probabilistic epigenesis prompt me to a reconsideration of the various roles of experience in individual development. I originally thought (Gottlieb, 1976a) there were three main ways experience (stimulation or activity) can influence development, as shown in Fig. 4.7. Inductive experiences are the most fundamental because they are essential to the achievement of a given developmental endpoint; if the inductive experiences do not occur, the endpoint is not achieved. English speakers speak English because they have been exposed to English speakers during the course of development: That is an example of an inductive effect of experience. Facilitating experiences hasten the temporal appearance of an endpoint, lower latencies, or lower thresholds of responsiveness; they affect the quantitative aspects of development, operating in conjunction with inductive experiences. Maintenance refers to the necessary role of experience to keep already achieved, induced endpoints functional; without such experiences, already developed endpoints will decay or be lost. These three roles of experience apply to both the behavioral and neural levels of analysis (Gottlieb, 1976b).

I initially formulated the three roles of experience or, as seems preferable to most neuroscientists, activity, at the behavioral and neural levels of analysis in two articles published in 1976. After that time, I discovered that experience can also play a canalizing role in development (Gottlieb, 1991a), as will be described in detail in chapter 6. A discussion of experiential canalization is pertinent to an understanding of the experimental results described in this chapter and the results described in the next chapter (5), as well as chapter 6.

Canalization is a narrowing of responsiveness as a consequence of

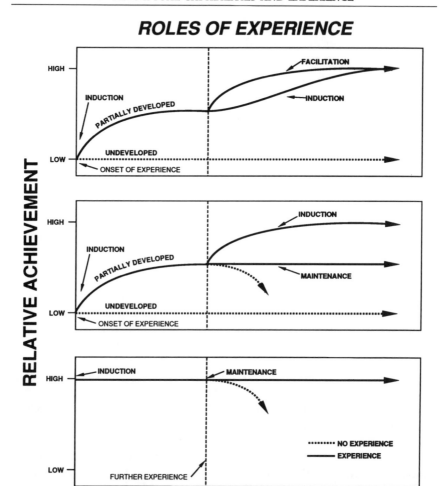

FIG. 4.7.   Three main roles of experience: induction, facilitation, maintenance. (Two subtypes of induction, canalization and malleability, are described in text.)

experience. That is, in its "initial state" on day 22, the aurally inexperienced embryo's perceptual system is responsive to a wide range of repetition rates (2.8–5.5 notes/sec). As a consequence of experiencing its contact call, that range of responsiveness is narrowed down (deleting responsiveness at the lower reaches of the range), whereas in the absence of experiencing its contact call, the range is not only not contracted but expands to include 2.3 notes/sec at the lower end, as shown by the devocalized mallard ducklings' equal responsiveness to the mallard call at 2.3 notes/sec and 3.7 notes/sec in the previous chapter.

This experiential canalization process is very similar to what happens in the developing nervous system: The initially "exuberant" (i.e., very large) number of synaptic contacts is pruned by experience. This is a cardinal feature of Gerald Edelman's (1987) notion of "neural Darwinism." Activity leads to the retention of synapses, and inactivity leads to their dissolution (Changeux & Danchin, 1976). At the psychological level of analysis with respect to speech perception, as an example, during the first year of postnatal life human infants are responsive to the universal range of phonemes that occur in all languages, but by the end of their first year they are responsive only to the phonemes of their native language, that is, the phonemes that they have experienced as native listeners and producers (Werker & Tees, 1984).

The canalization process is sometimes misconstrued as demonstrating the maintenance role of experience, which overlooks the important fact that canalizing experiences bring about a narrowing or enhanced specificity of responsiveness, and the absence of experience, as in the case of the devocalized duck embryo, leads to the continuance or enlargement of the initial broad range of responsiveness. The positive maintenance role of experience comes into play only after the specificity of synaptic contacts or the selective responsiveness to phonemes is in place. Thus, I would now add canalization as a fourth role of experience, with experience being defined broadly to signify the contribution of *functional activity* at the behavioral and neural levels of analysis, whether the activity arises from external or internal sources.

In this context, my more recent research has revealed yet a fifth role of experience, experiences that lead to an enhanced malleability or plasticity (the opposite of the narrowing of responsiveness brought about by canalizing experiences). In ducklings, as we shall see in chapter 7, social rearing can have a profound inductive effect on malleability (Gottlieb, 1991b, 1993; Sexton, 1994).

Because canalization and malleability require inductive experiences, perhaps they should be classified as subtypes under induction. In later chapters I shall present experiments bearing on canalization and malleability, but first it is necessary to back up a little to present experimental evidence that the finding of a prenatal experiential canalization of instinctive behavior in mallard ducklings is not an unusual or singular phenomenon in the field of animal behavior.

5

Experimental Demonstration of Probabilistic Epigenesis in Wood Ducklings

The hole-nesting wood duckling, like the ground-nesting mallard duckling, is able to identify the maternal assembly call of its species in advance of exposure to a wood duck hen or her call. For maternally inexperienced mallard ducklings and wood ducklings to be able to discern the assembly call of their respective species, there must be particular acoustic features that invariably occur in the call, ones to which the young bird is always attracted. The developmental investigation of the wood duckling's ability to identify a maternal call of its species took the same form as the investigation in the mallard duckling: (a) determine the critical acoustic features of the wood duck maternal call, and (b) attempt to link those features to the wood duck embryo or hatchling's early auditory experience (exposure to its own and sib vocalizations). Once again, the question is whether the critical acoustic features of the wood duck maternal call are somehow represented in the embryonic or neonatal vocalizations.

KEY ACOUSTIC FEATURES OF WOOD DUCK MATERNAL CALL

The rationale for determining the key acoustic features of the wood duck maternal call was as follows. The first step was to record the maternal calls of 11 wood duck hens as they called their young from nests in the field and subject these calls to an intensive acoustic analysis (Miller & Gottlieb, 1976). The acoustic features that showed little variation across hens would be the prime candidates for the critical perceptual features of the call used by maternally naive wood ducklings. That is, if the ducklings somehow are uniquely sensitive to the acoustic features of the generalized maternal call of their species in advance of exposure to a hen, these features would have to be rather stably represented in the calls of individual hens; otherwise, the ducklings' manifest ability to identify a maternal call of their species in advance of exposure to it would not be in evidence, as we know it is.

Repetition Rate

The field study of the maternal calls of 11 wood duck hens indicated that the frequency modulation of the notes and the repetition rates of the calls were the most stable features across hens (Miller & Gottlieb, 1976). To examine the importance of repetition rate, a wood duck maternal call (Fig. 5.1) was altered by splicing blank pieces of tape between the notes to decrease the repetition rate of the call from 6.9 notes/sec to 4.0 notes/sec. Because the only feature altered was repetition rate, a preference shown by maternally naive birds for the 6.9 over the 4.0 rate in a simultaneous auditory choice test would indicate that repetition rate is one of the key perceptual features of the wood duck maternal call.

As can be seen in Fig. 5.2, in a simultaneous choice test, the 14 maternally naive wood ducklings that made a choice did not show a preference for the normal (6.9) over the slowed (4.0) maternal call. Because the particular maternal call selected for this test was a bit slower than the modal repetition rate (8.8 notes/sec) of other hens recorded in the field (Miller & Gottlieb, 1976), an additional group of wood ducklings was tested with the same call at 6.9 notes/sec and a faster version of that call at 8.8 notes/sec.[1] (The faster version was made by deleting blank space between the notes, as shown in Fig. 5.1.) The results of that test are shown in Fig. 5.2. Again, the 16 birds that made a choice did not prefer either repetition rate, that of the normal call (6.9), or the modal rate for the species (8.8). So, although repetition rate may not be completely unimportant, it is clearly not a key perceptual feature of the wood duck maternal call for maternally naive wood ducklings.

Frequency Modulation

Because the frequency modulation (FM) of the notes of the wood duck maternal call was rather stable across the 11 wood duck hens that Miller and

---

[1]The figures for repetition rates given in this sentence are slightly different from those that appeared in Miller and Gottlieb (1976) and Gottlieb (1974), because of a difference in method of calculation. The present figures have been calculated by a more appropriate method (Scoville & Gottlieb, 1978).

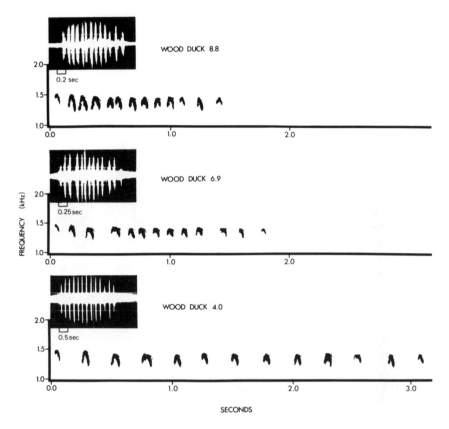

FIG. 5.1. Scale-magnified Sonagrams of wood duck maternal call at three different repetition rates. To speed or slow the normal call (6.9 notes/sec), it was necessary merely to delete space between the notes (8.8 notes/sec), or to add blank pieces of recording tape between the notes (4.0 notes/sec).

I (Miller & Gottlieb, 1976) recorded in the field, that was the next acoustic feature to be examined in the laboratory experiments. As can be seen in Fig. 5.3, we classified 661 notes from the 11 hens as primarily descending, primarily ascending, or symmetrically ascending–descending in FM, and 70% of the notes were mostly descending in modulation. All of the hens uttered descending notes, and all but one uttered many more descending notes than ascending or symmetrically ascending–descending notes (Fig. 4 in Miller & Gottlieb, 1976). The one exception uttered an almost equal number of descending and ascending notes.

If the descending FM of the notes is a critical perceptual feature, it should be possible to synthesize (fabricate) a call that is as attractive to wood ducklings as a natural call. To determine if the descending feature of the FM is indeed the critical aspect, maternally naive wood ducklings were

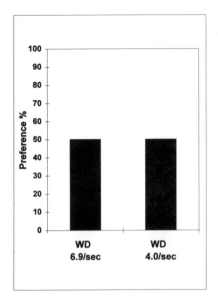 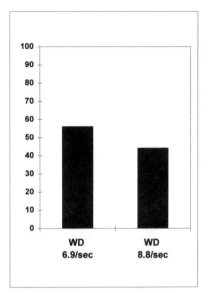

FIG. 5.2.   Preferences of wood ducklings in simultaneous auditory choice tests with wood duck maternal call at three different repetition rates: 8.8 notes/sec (species mode), 6.9 notes/sec, and 4.0 notes/sec. (Results for 6.9 vs. 4.0 from Gottlieb, 1974, Table 5; 8.8 vs. 6.9 from Gottlieb, 1981, Fig. 1.3.)

FREQUENCY MODULATION

DESCENDING

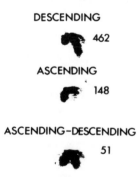

ASCENDING

ASCENDING–DESCENDING

FIG. 5.3. Tally of frequency modulation of 661 notes in maternal calls from 11 wood duck hens recorded during exodus from nest in the field. Of all the notes, 70% had a pronounced descending modulation (462). (Data abstracted from Miller & Gottlieb, 1976, Fig. 4.)

tested for their preference among three synthetic calls: descending, ascending, and ascending–descending. It is important to realize that the call notes differed only in their FM; otherwise the calls were similar in frequency range, note length, burst length, and repetition rate (Fig. 5.4). As is evident in Fig. 5.4, each of the synthetic calls has an ascending and a descending component, with the ascending–descending call being closest to symmetrical

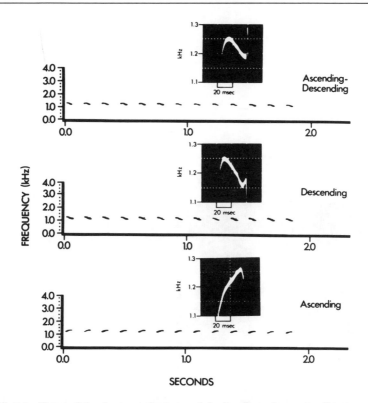

FIG. 5.4. Notes of the three synthetic wood duck calls to determine if maternally naive wood ducklings have a preference for the modal frequency modulation (descending). Each of the three calls contained 13 notes with a repetition rate of 6.9/sec to be equivalent in those respects to the natural wood duck call shown in Fig. 5.5. (Although the duration of notes in the natural wood duck call varies from 50 to 75 msec, all notes in the synthetic calls are a standard length.)

in that respect, the descending call having a more pronounced descending component, and the ascending call having a greater ascending component.

As shown in Table 5.1, in simultaneous auditory choice tests, maternally naive wood ducklings preferred the synthetic descending call to the other synthetic calls. The next step in determining whether the descending synthetic call contained the critical components of the natural wood duck maternal call was to place the synthetic descending call in opposition to the natural wood duck maternal call in a simultaneous auditory choice test. The natural call was the one used earlier in the repetition-rate experiments. The detailed FM of the notes of the natural call is shown in Fig. 5.5, which can be contrasted to the descending synthetic call in Fig. 5.4. Both calls contained 13 notes and had a repetition rate of 6.9 notes/sec. If the synthetic call contained the critical perceptual components that underlie the

TABLE 5.1
Preference of Wood Ducklings in Simultaneous Auditory Choice Tests With
Descending, Ascending, and Ascending-Descending Synthetic Calls

Synthetic Calls	Preference		
	Descending	Ascending	Ascending-Descending
Descending vs. ascending	13*	5	—
Descending vs. ascending–descending	20**	—	5
Ascending vs. ascending–descending	—	8	15

*$p = .05$, **$p < .01$, binomial test. (Data from Gottlieb, 1974, Table 1.)

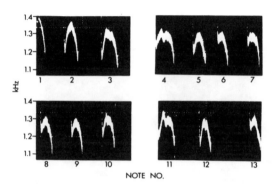

FIG. 5.5. Davis frequency chronogram of natural wood duck maternal call used in simultaneous choice tests with synthetic calls. Note that virtually all the notes of the natural call have a more pronounced descending component. Horizontal time scale = 50 msec/division. (Compare with synthetic calls in Fig. 5.4.)

attractiveness of the natural call, the ducklings should find the synthetic call as attractive as the natural call in the choice test. Otherwise, the birds would prefer the natural call.

As shown in Fig. 5.6, maternally naive wood ducklings did not show a preference between the descending synthetic call and the natural wood duck call. To determine the specificity of the perceptual selectivity of the ducklings, another group of wood ducklings was tested with the ascending-descending synthetic call versus the natural call. In that case, the ducklings preferred the natural call to the synthetic one (Fig. 5.6). Therefore, the descending FM is a critical perceptual feature of the wood duck maternal call.

EFFECT OF AUDITORY DEPRIVATION ON SPECIES-SPECIFIC PERCEPTUAL PREFERENCE: DESCENDING FREQUENCY MODULATION

We are now in a position to ask what role, if any, prior auditory experience plays in the development of the ducklings' selective response to the

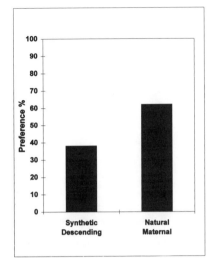
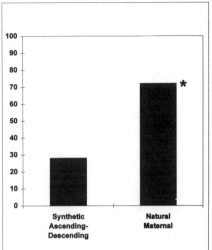

FIG. 5.6.   Preference of wood ducklings in simultaneous auditory choice tests with synthetic versus natural maternal calls. (Asterisk signifies statistically reliable difference.)

descending frequency modulation of the wood duck maternal call. According to the results just presented, maternally naive wood ducklings identify the call of a hen of their species primarily on the basis of its FM, with the descending aspect being the pertinent cue. Given the small differences in steepness and depth of descent between the synthetic descending call and the synthetic ascending–descending call used in the present experiments (Fig. 5.4), the manifest preference of the wood ducklings for the former in a simultaneous choice test attests not only to the importance of the descending modulation but to the fineness of the duckling's discriminative ability. The only vocal–auditory experience the birds had before being tested was hearing their own and sib vocalizations. Could this experience play a role in the development of their very keen perception of a key feature of the maternal call?

As in many other species, the embryonic and neonatal wood duckling is capable of uttering at least two acoustically distinct vocalizations: contact-contentment and alarm-distress calls. (As documented in chapter 3, when the embryos move into the air space at the large end of the egg a few days before hatching, they begin breathing and thus can vocalize before hatching [Gottlieb & Vandenbergh, 1968].) As shown in Fig. 5.7, the notes of both the alarm-distress and contact-contentment vocalizations exhibit significant descending frequency modulations. In the case of the alarm call, the descending component traverses almost 3000 Hz (from about 6000 to almost 3000 Hz).

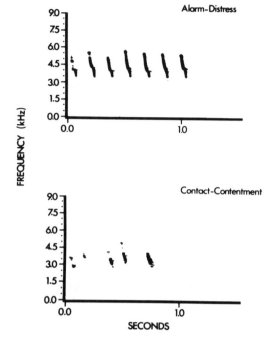

FREQUENCY (kHz)

SECONDS

FIG. 5.7. Sonagrams of alarm-distress and contact-contentment calls of wood ducklings. Notice the pronounced descending frequency modulation of each call extending over several thousand hertz. Notes in these calls are not in the same frequency range of the wood duck maternal call (compare Sonagrams in Fig. 5.8).

The descending FM of the contact-contentment call goes from about 4500 to 2500 Hz. Both of these modulations are considerably outside the range of the fundamental frequency of the maternal call, which is around 1300 Hz (Fig. 5.5). Could the birds be "abstracting" the descending FM from exposure to their own and sib vocalizations? The first step in answering that question would be to deprive the birds of hearing their own and sib vocalizations via embryonic devocalization, housing them in soundproof incubator compartments, and then testing them, for example, with the descending versus ascending synthetic calls. If hearing the embryonic and neonatal vocalizations plays no role in the development of their preference for the descending FM, then such aurally deprived birds would prefer the descending call over the ascending one. If, however, exposure to the embryonic and neonatal vocalizations does play a significant role in the development of their perceptual preference, then aurally deprived wood ducklings might not show the normal preference for the descending call. That is the investigative strategy that I followed (Gottlieb, 1980b), with the exception of not muting the birds as embryos. Wood ducklings do not tolerate the muting operation as well as mallard ducklings, so we merely placed the birds in individual auditory isolation as embryos and tested them after hatching. Admittedly, this is not as precise a deprivation procedure as is devocalization, but it does work with wood ducklings.

As can be seen in Table 5.2, the aurally isolated wood ducklings did not show the normal preference for the descending synthetic call.

These results indicate that a certain amount of exposure to their own and/or conspecific vocalizations plays some role in the development of the wood duckling's species-specific perceptual preference for the descending FM characteristic of the wood duck maternal call. One of the roles normal experience could play is temporal regulation or facilitation; that is, it may facilitate the appearance of the usual degree of perceptual specificity by a customary age. In that case, the absence of such experience would merely lead to a delay in the appearance of the ability, as observed in the mallard ducklings' response to the high-frequency components of the maternal call of its species: Isolates did not show the usual responsiveness to the high-frequency components at 24 hr after hatching, but they did show such responsiveness at 48 hr after hatching (chapter 3).

To determine if auditory deprivation merely slows the rate of normal perceptual development in wood ducklings, aurally isolated wood ducklings were tested in the next experiment with the descending versus ascending calls at 48 hr after hatching instead of at 24 hr as in the previous experiment.

As shown in Table 5.2, the aurally isolated wood ducklings did not show the preference for the descending FM when tested at 48 hr after hatching. Although it is possible that delaying testing beyond 48 hr might show the eventual appearance of the preference for the descending FM, in the previous instance where this was investigated (chapter 3), one in which there was improvement at 48 hr, there was a subsequent deterioration when the isolates were tested for the first time at 65 hr after hatching. In an unpublished experiment with wood ducklings related to this study, I examined the behavior of aural isolates 65 hr after hatching; their responsiveness in the test declined from 73% at 48 hr to 42% at 65 hr (and

TABLE 5.2
Aurally Isolated and Normal Communal Wood Ducklings' Preference in Simultaneous
Auditory Choice Test With Descending Versus Ascending Synthetic Calls

Prior Auditory Experience	Age at Testing	Preference	
		Descending	Ascending
Normal communal	24hr	13*	5
Auditory isolation	24hr	16	14
Auditory isolation	48hr	20	15

*p < .05, binomial test. Data from Gottlieb (1980b, Tables 1 and 2).

there was no evidence of improvement). Consequently, I did not pursue the investigation beyond 48 hr with the wood ducklings in the present study.

SPECIFICITY OF EXPERIENTIAL REQUIREMENT TO MAINTAIN OR INDUCE SPECIES-SPECIFIC FREQUENCY MODULATION PREFERENCE

Although the wood ducklings show a preference for the descending call when they are incubated and brooded communally, the question is whether it is exposure to their own and conspecific vocalizations that is specifically responsible for the preference. To examine this question, wood duck embryos were placed in individual auditory isolation, as in the previous experiments, and exposed to a repeated recording of a single burst of a wood duckling distress call (Fig. 5.8) for 5 min every hour until they were tested at 24 hr with the descending versus ascending calls. I chose to expose them to the alarm-distress call rather than to the contact-contentment call because I did not think a positive emotional connotation or association was necessarily required if such experience were to be effective, as has been theorized by Guyomarc'h (1972) in his extension of Schneirla's theory (1965). Because both the alarm-distress and contact-contentment calls are markedly descending in FM, I thought it possible that one call might be just as effective as the other. In contrast to the mallards, which only rarely make distress calls, wood ducklings are extremely excitable and make a large number of alarm calls, so it is a call they are exposed to under normal conditions.

It is significant to note that the embryos and hatchlings spend a great deal

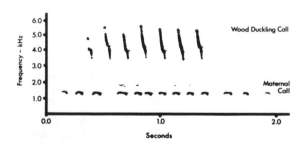

FIG. 5.8. Seven-note burst of wood duckling alarm-distress call (top). Aurally isolated wood ducklings were played this call in either the forward direction (descending modulation) or backward direction (ascending modulation) prior to testing to determine if such exposure would influence their choice of the descending versus ascending synthetic calls. Frequency range of maternal call is shown at bottom for comparison.

of time in sleep during the period of stimulation, and that the stimulation is of course not contingent on their state or activity. In mallards, in which the eggs are plentiful and easy to obtain, it was possible to show that 5 min per hour of relevant stimulation confined to the embryonic period remained effective up to 65 hr after hatching, the latest the birds were tested (chapter 3). Because wood duck eggs are neither plentiful nor easy to acquire (see Fig. 5.9), parametric study is not practical and I elected to continue the stimulation up to the time of testing rather than to terminate it around the time of hatching.

As can be seen in Table 5.3, the isolates stimulated with the alarm-distress call showed the normal preference for the descending synthetic call.

To determine if the normal preference for the descending synthetic maternal call is specifically a consequence of the birds experiencing the descending FM of the duckling alarm-distress call, in the next experiment isolated ducklings were exposed to the same alarm-distress call played backward (making it an ascending FM instead of a descending FM) on the same schedule as before and tested as before with the ascending versus descending synthetic calls at 24 hr after hatching.

There are several possible outcomes in the present experiment. If the experiential requirement is not specific, the alarm call played backward would be sufficient for the manifestation of the normal preference for the descending synthetic maternal call. If the experiential requirement is specific, as it is in the mallards, the wood ducklings in the present experiment would not show the normal preference. Furthermore, if the developing perceptual preference is relatively unconstrained in the direction of its development, stimulation with the alarm call in the backward direction could lead to a preference for the ascending synthetic call over the descending one.

As shown in Table 5.3, the isolates stimulated with the alarm-distress call in the backward direction did not show the normal preference for the

TABLE 5.3
Wood Ducklings' Preference in Simultaneous Auditory Choice Tests With Descending Versus Ascending Synthetic Calls at 24 hr After Hatching

	Preference	
Prior Auditory Experience	Descending	Ascending
Auditory isolation	16	14
Auditory isolation stimulated with alarm-distress call	17*	7
Auditory isolation stimulated with alarm-distress call played backward	13	19
Normal communal	13*	5

*$p < .05$, binomial test. Data from Gottlieb (1980b, Table 3).

FIG. 5.9. The late T. C. Schneirla steadies the ladder as the author prepares to collect wild wood duck eggs from an artificial hole-nest at the Dorothea Dix Animal Behavior Field Station in 1962.

descending synthetic call, nor did they show a reliable preference for the ascending call.

There are several especially significant features in the present results. First and foremost, the development of the normal preference for a critical acoustic feature of the maternal call of the species is a highly specific consequence of the embryo and hatchling hearing its own and/or sib alarm-distress calls. Because the alarm-distress call is in a completely different frequency range than the maternal call, and its FM extends over a far greater range than the maternal call, the birds would appear to be "abstracting" the descending frequency modulation component when they subsequently respond to it in the maternal call (see Fig. 5.8).

The isolated ducklings did not develop a preference for the ascending frequency modulation when exposed to the duckling alarm-distress call played backward, thus suggesting a developmental constraint in the malleability of the FM preference. Because the isolated ducklings were not mute, at present we cannot rule out the interesting possibility that this constraint may be due to their hearing their own contentment and distress vocalizations (both having a descending FM), which could have buffered them against the effect of exposure to the ascending FM (distress call played backward). The other possibility is that the constraint stems entirely from endogenous auditory neuroanatomical and neurophysiological limiting factors that simply preclude the development of a preference for ascending FM at the developmental stage at which the experiments were conducted.

INEFFECTIVENESS OF SELF-STIMULATION IN WOOD DUCKLINGS

Because wood ducklings vocalize copiously while in individual auditory isolation, it seems as if experiencing their own vocalizations is ineffective in producing the normal preference for the descending frequency modulation of the notes of the maternal call. It could be, however, that when the isolates are stimulated with the recording of a wood duckling distress call they produce more descending notes themselves than when unstimulated, so enhanced self-stimulation could be involved in a subtle and indirect way. To test this possibility, we (a team effort as always) elected to undertake the arduous task of monitoring the vocalizations of stimulated and unstimulated wood duck embryos and hatchlings up until they were tested at 24 hr after hatching.

Because wild wood ducks only lay eggs during the spring, it took three seasons to collect enough eggs to complete this experiment. During the course of the experiment, we recorded a sample size of 1,237 notes in the unstimulated group and 1,418 notes in the stimulated group. Three experienced raters then independently scored the FM of each of the notes as (a) primarily ascending, (b) primarily descending, (c) symmetrical (or nearly so), or (d) shape of FM doubtful. If two of the three raters agreed on the character of a note, the note was retained in the sample. Agreement was very high: Only 25 of the total sample of 2,655 notes had to be excluded from the final tally.

In order to be able to relate each duckling's vocal output to its preference for the descending or ascending FM of the synthetic maternal calls, we tested all of the birds, thereby also replicating our previous result: The unstimulated group showed no preference, whereas the stimulated group preferred the descending FM (Gottlieb, 1984).

To pursue the analysis of self-stimulation and preference for the descending maternal call in the test, the vocal activity of the birds was analyzed in several different ways. First, Fig. 5.10 presents the overall distribution of the FMs of all the notes produced by the birds. The distributions are similar in the unstimulated and stimulated groups, with the exception that the stimulated birds produced more *ascending* notes (more on this point later). There is no evidence here that the stimulated birds produced more descending notes than the unstimulated birds. However, the most relevant question is whether the birds that preferred the descending maternal call produced more descending notes than the birds that did not prefer the descending call: The answer is negative in both the stimulated and unstimulated groups. A further question is whether the birds that preferred the descending call were more vocal (all types) than the other birds, and that answer is also negative in both groups.

A final way to analyze these data is to examine the stimulated birds' vocal reactivity during the periods when the tape-recorded distress call was being

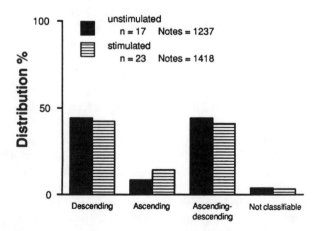

Distribution % of FM Notes Produced
by Isolated Wood Ducklings

FIG. 5.10. Distribution of frequency-modulated notes produced by stimulated and unstimulated wood ducklings while in individual isolation.

broadcast. In terms of change from prestimulation baseline, the stimulated birds produced fewer descending calls and more ascending calls during the broadcast of the descending sib call. Could it be that by producing ascending notes during the playback the stimulated birds were sharpening their perception of descending and ascending calls, and thus were able to perform better in the test with the descending and ascending maternal calls? To answer that question, the production of ascending notes by the birds that preferred the descending maternal call was compared with that of the other stimulated birds that responded in the test, and no difference was found.

In summary, although the stimulated birds produced more ascending notes than the unstimulated birds, no differences were found in the overall vocal behavior, vocal reactivity, or specific kinds of FM produced by the birds that preferred the descending maternal call and the other birds that responded in the choice test. Thus, auditory self-stimulation apparently plays no role in the development of the wood duckling's usual species-specific perceptual preference for the characteristically descending frequency modulation of the wood duck maternal call. Rather, wood ducklings must hear the descending call note of a sib in order to show the species-specific preference. This is in marked contrast to mallard ducklings, in which a very small amount of strictly embryonic auditory self-stimulation suffices to induce/canalize their selective response to the mallard maternal call (delayed devocalization experiment described in chapter 3).

In chapter 7 we will explore whether the wood duckling's greater dependence on the socially produced acoustic environment signifies that it is more acoustically malleable than the mallard duckling.

ROLE OF EXPERIENCE

What role is experience playing in the present case? Because the isolated ducklings do not manifest the normal preference for a descending frequency modulation even if tested at a later age, experience is not playing a facilitative role. That is, the isolates are not merely showing a lag in perceptual development as a consequence of experiential deprivation. Therefore, experience is serving either an inductive or a maintenance function. That issue can be decided by examining the response of embryos to the descending and ascending FMs in advance of normal auditory experience. If the embryos are responsive to both the descending and ascending FM calls, then auditory experience is inducing the normal postnatal preference for the descending call. If the embryos are responsive to the descending FM call and not to the ascending call, then auditory experience is serving a maintenance function.

As noted in chapters 3 and 4, in the mallard ducklings, the exposure to their embryonic call plays an inducing/canalizing role, because, in advance of auditory experience, the mallard duck embryo is responsive to a wider range of repetition rates than is the aurally experienced embryo and hatchling. The same is true in the wood duckling: The aurally inexperienced wood duck embryo is responsive to both ascending and descending FM, and becomes unresponsive to ascending notes after auditory experience (Gottlieb, 1983). In the absence of hearing sib vocalizations, the wood duckling remains responsive to ascending notes. Thus, in the wood duckling, the experience of hearing sib vocalizations plays an inducing/canalizing role as it does in mallard ducklings.

NONOBVIOUS EXPERIENTIAL BASIS OF UNLEARNED BEHAVIOR

As one examines the acoustic features of the wood duck maternal call and the wood duckling distress call (Figs. 5.5, 5.7, and 5.8), it is far from obvious that the distress call could provide indispensable patterned "information" for the embryo and hatchling concerning a critical acoustic feature of the species' maternal call. If one were to make such an hypothesis prior to the findings presented here and in chapter 3, it would seem not only nonobvious but farfetched—superficially, the calls do not seem to share *any* common features. The invitation to search for nonobvious experiential bases of instinctive behavior, implicit and explicit in the writings of Kuo (1976), Schneirla (1956), and Lehrman (1953), has been largely ignored. As de Santillana has written, "But nothing is so easy to ignore as something that does not yield freely to understanding" (de Santillana & von Dechend, 1977, p. xii).

The finding of nonobvious experiential bases of unlearned behavior forces us to think in a new way about the role of experience in the development of behavior that is thought of as instinctive. As discussed in chapters 3 and 4, there is not only one role of experience (conditioning or traditional forms of learning) during development, there are at least three others: induction, facilitation, and maintenance. The interesting things about these three modes are that (a) they do not fit the definition of traditional (i.e., associationistic) learning, (b) they entail normally occurring specific patterns of stimulation to achieve the species-specific behavioral outcomes, and (c) their role is not necessarily developmentally obvious (i.e., they are not related to their outcomes in a straightforward way).

The search for nonobvious experiential bases to instinctive behavior is in line with evolutionary considerations. Natural selection works on adaptive behavioral phenotypes; it is completely indifferent to the particular pathway taken by the phenotype during the course of development (Gottlieb, 1966; Lehrman, 1970; Mivart, 1871). Because all forms (pathways) of development involve genetic activity, natural selection need not favor one developmental pathway over another in the ontogenesis of unlearned behavior. As mentioned before, natural selection involves a selection for the entire developmental manifold, including both the organic and normally occurring stimulative features of ontogeny. Thus, nonobvious experiential involvement of a patterned kind may be much more widespread than heretofore realized. Only developmental investigations of unlearned behavior can answer that question, and that has not been a popular form of experimentation, because influential scientists have looked on seemingly innate or unlearned behavior as a predetermined outcome of development, not as a probabilistic epigenetic phenomenon as, for example, David B. Miller (1994), Robert Lickliter and Heather Banker (1994), and Thomas Bidell and Kurt Fischer (1997) have in their research.

SUMMARY AND CONCLUSIONS

The experimental results reviewed in this chapter and in chapter 3 support the notion that normal embryonic or prenatal experience plays an essential role in the development of species-specific behavior after birth or hatching (behavior otherwise thought of as instinctive or innate). In this century, Kuo (1921) was the first to raise the possibility that features of instinctive behavior of neonates may be a consequence of experience that occurred in the embryo and, building on Kuo's thoughts and the hypotheses stemming from his research, Lehrman (1953, 1970) and Schneirla (1956, 1965) kept the idea alive, although it was admittedly empirically unfounded. I regret that my scholarly predecessors, each of whom I knew personally, passed away before the completion of this program of research. Their ideas stimulated acrimonious retorts from those strongly inclined to predetermination, and it is satisfying to have been able to provide experimental vindication of their point of view. Experiments are often said to be the cornerstone of science, but it is ideas that motivate the experiments, and good ideas are hard to come by.

When I was a graduate student and hungrily read W. H. Thorpe's (1956) *Learning and Instinct in Animals*, I was so distressed about the dismissive way he characterized the possible prenatal antecedents to postnatal behavior that I was never able to rid my mind of what he wrote in the first, and repeated in the second, edition (1963) of his book:

The far-fetched suggestion of Lehrman (1953), based on the observations of Kuo (1932), that the pecking behavior is itself "learnt" in the embryo, as a result of tactile stimulation (via the yolk sac) of the amniotic pulsations, hardly seems to bear close consideration. (Thorpe, 1963, p. 346)

In the same work, on page 351, Thorpe wrote in greater detail:

It has been pointed out (Kuo, 1932) that practically every physiological organ of the birds [sic] is in functional condition before hatching, and that practically every part of the muscular system of the young bird has been exercised some days before the end of the first half of the incubation period. This of course is true enough, and these embryonic movements are the elements out of which many of the responses shown by the bird after hatching are built; but it is a misapprehension to suppose, as Kuo does, that learning in any sense is taking place in the embryonic state. What is happening inside the egg when we detect movements of the embryo is, mainly if not entirely, a process of maturation of the innate behaviour pattern.

The point that Kuo was trying to make is that exercise of the motor movements may be an important contributor to their maturation. The way Kuo wrote about this in 1932 and Lehrman wrote about it in 1953 seems also to have been misinterpreted by Lorenz (1965, p. 104) as "the chick learning to peck inside the egg." I think this sort of misunderstanding by outstanding figures in the field stems, at least in part, from holding a strictly dichotomous view on sources of influence: Development is a consequence either of maturation or of learning. It was to try to circumvent this problem that I (and Schneirla before me) opted for the more inclusive term *experience*, which includes learning but goes beyond it to encompass endogenous as well as exogenous sensory stimulation, motor activity, and neural activity as crucial contributors to species-specific or normal behavioral development. As I noted in chapter 4, when the receptors on the surface of a nerve cell are excited or inhibited, I do not think it makes any difference to the receptors whether the source of stimulation was exogenous or endogenous to the organism.

6

Experiential Canalization of Species-Specific Behavior

In addition to the way I defined canalization in chapter 4, it has been utilized in several related ways in the psychological literature. The concept of canalization was originally put forward by E. B. Holt (1931) to call attention to prenatal conditioning (learning) as a factor in narrowing down the initially diffuse or random nature of motor activity in the embryo or fetus. Holt saw the motor activity of the fetus and the infant as becoming organized through spatial and temporal contiguity learning, which narrowed down originally diffuse neural pathways to a definite neural reflex arc. So, for Holt, canalization referred to the development of specific sensorimotor pathways out of an original multiplicity of such pathways — it was a label to capture the developmental–behavioral phenomenon of progression from diffuse to ordered or organized motor activity, with contiguity conditioning supposedly serving as the experiential mechanism whereby organized motor activity was achieved. Although Holt's concept of neural reflex circles to account for the development of grasping in the fetus and infant was an ingenious example of his point of view, it remained an entirely speculative theoretical solution of primarily historical interest. Subsequent empirical studies in behavioral embryology kept open the possibility that motor movements in the embryo and fetus of some species are patterned from the start, whereas in others they appear to be random early and patterned later (see reviews in Gottlieb, 1970; Hamburger, 1973; Oppenheim, 1974).

Another meaning of canalization was put forward by Z.-Y. Kuo (1976). For Kuo, canalization was a broadly applicable principle that says as development proceeds the originally great range of behavioral potentials or plasticity narrows, signifying that the range of possibilities of behavioral development always exceeds the range of behavior that is actualized during the course of individual development. The channeling of behavior, and the correlated decrease of plasticity, over the course of ontogenesis is to be explained by the individual's particular developmental history, which, for Kuo, included biochemistry, physiology, and anatomy, as well as experience. Thus, Kuo's principle of canalization is a much broader idea than Holt's, and its appropriately descriptive intent would seem to mesh, in the most general terms, with almost everyone's idea of what happens during individual development. As with Holt, Kuo did not actually provide any explicit empirical or experimental support for what seems a face-valid proposition to many developmental psychologists and psychobiologists.

More recently, a number of psychologists have picked up on another, very different concept of canalization. That is the developmental geneticist Waddington's (1942) notion that early normal or species-typical physiological and anatomical development can withstand great assaults or perturbations and still return to (or remain on) its usual developmental pathway, thus producing the usual or normal phenotype. Waddington's concept of canalization says that usual developmental pathways are so strongly buffered (by genes — Waddington, 1957, p. 36, Fig. 5) that normal or species-typical development can be only temporarily derailed. Waddington (1968) used the term *chreod* to express his idea more succinctly: A chreod, according to Waddington, is a "fated" or predetermined developmental pathway. Thus, for Waddington, embryonic anatomical and physiological development consists of a number of highly buffered (i.e., virtually fixed) endogenous pathways. Once again, the idea seems so face-valid that no empirical support is presented for it and the potential mechanism itself is described in only the most figurative or metaphorical terms. Individual development is characterized as a ball rolling down valleys of greater or lesser height that are said to make up the geography of the "epigenetic landscape" (Waddington, 1942, 1953, 1957, 1968).

Because the epigenetic landscape in which Waddington's developmental pathways are embedded is merely figurative or metaphorical, his concept of canalization is devoid of any empirical content and thus supplies us with no concrete understanding or hypotheses concerning the developmental process that is involved. The appeal of Waddington's notion, despite its conceptual and mechanistic emptiness, is testified to by its wide adoption in various developmental–psychological models as a virtual synonym for what was previously called the innate, native, or maturational component in behavioral development (e.g., Fishbein, 1976; Kovach & Wilson, 1988; Lumsden & Wilson, 1980; Parker & Gibson, 1979; Scarr-Salapatek, 1976). The strictly genetic determination of canalization and its connection to the reaction range are captured in the following quotations:

Canalization is the genetic limitation of phenotypic development to a few possible phenotypes rather than an infinite variety . . . . The genetic restriction of possible phenotypes is a result of coadapted gene complexes (sets of genes that have evolved together) that buffer the developmental pattern against deviation outside of the normal, species range of variation . . . .

The limitation of possible phenotypes to a few rather than many is a fact of genotypes' ranges of reaction. This concept can best be expressed developmentally in the notion of developmental pathways, which Waddington calls "creods." [sic] Behavioral creods are essentially similar to Piaget's notion of schemes, organized patterns of behavior that develop in characteristic ways. They represent the biases the organism has toward acquiring some rather than other forms of behavior . . . .

These biases are undoubtedly genotypic. (Scarr-Salapatek, 1976, p. 63)

Waddington himself expressed the genetic determination of canalization as follows:

The epigenetic feed-back mechanisms on which canalization depends can, of course, be regarded as examples of gene interaction. Interaction between two allelomorphs is referred to by such terms as dominance, recessiveness, overdominance, etc. Interactions between different loci come under the heading of epistasis. This is perhaps most usually thought of in terms of interaction between only two or three loci. We know, however, that in the development of any one organ very many genes may be involved, and in canalized epigenetic systems we are probably confronted with interactions between comparatively large numbers of genes. (Waddington, 1957, p. 131)

Fourteen years later, Waddington reiterated the strictly genetic determination of canalization as follows:

The degree to which each pathway is canalized or self-establishing is dependent on the particular alleles of the genes involved in it; and it can be altered by selection of a population either for alleles which fit better into the canalized system (and thus increase the organism's resistance of modification) or for alleles which do not integrate so well with the others (and thus lend to decreased resistance to external influences). (Waddington, 1971, pp. 20–21)[1]

CONTEMPORARY DEVELOPMENTAL THEORY

In recent years, what might be called a "systems view" of individual development has been slowly catching on, both in biology and in psychol-

---

[1]I feel somewhat ungrateful showing Waddington to be a rather surprisingly strict genetic determinist by quoting his own words. He is one of the few major thinkers in evolutionary biology who felt it was necessary to take into account developmental and organismic considerations, whereas most of the major figures in the making of modern synthesis (neo-Darwinism) consider only natural selection and genetic variation to be the prime factors of importance in evolution.

ogy. The systems view sees individual development as hierarchically organized into multiple levels (e.g., genes, cytoplasm, cell, organ, organ system, organism, behavior, environment) that can mutually influence each other. The traffic is bidirectional, neither exclusively bottom-up or top-down. (A formal treatment of hierarchy theory can be found in Salthe, 1985, especially chapter 4.) Frances Degen Horowitz's (1987) recent review made the systems case for developmental psychology, and the geneticist Sewall Wright (1968) and the embryologists Ludwig von Bertalanffy (1933/1962) and Paul Weiss (1939/1969) have long been championing such a systems view for developmental genetics and developmental biology. The systems view includes developmental approaches and theories that have been called transactional (Dewey & Bentley, 1949; Sameroff, 1983), contextual (Lerner & Kaufman, 1985), interactive (Johnston, 1987), probabilistic epigenetic (Gottlieb, 1970), and individual-socioecological (Valsiner, 1987). The dynamic systems view of Thelen and Smith (1994), which comes out of chaos theory, is more of a theory (i.e., it has more specific predictive content) than the developmental systems metatheory that I am describing here. For the present purposes, I think the metatheoretical developmental-psychobiological systems view can be fairly represented by the schematic presented in Fig. 6.1.

The most important feature of the developmental systems view is the explicit recognition that the genes are an integral part of the system and their activity (i.e., genetic expression) is affected by events at other levels of the system, including the environment of the organism. It is a well-accepted fact, for example, that hormones circulating in the blood make their way into the cell and into the nucleus of the cell, where they activate DNA that results in the production of protein (Gorbman, Dickhoff, Vigna, Clark, & Ralph, 1983, p. 29, Fig. 1.13). The flow of hormones themselves can be affected by environmental events such as light, day length, nutrition,

BIDIRECTIONAL INFLUENCES

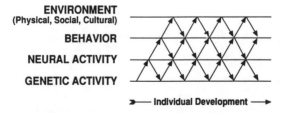

FIG. 6.1. The probabilistic epigenetic conceptual framework, showing a hierarchy of four mutually coacting levels of analysis in which there are "top-down" as well as "bottom-up" bidirectional influences. From *Individual Development and Evolution: The Genesis of Novel Behavior* by Gilbert Gottlieb. Copyright © 1991 by Oxford University Press, Inc. Reprinted by permission.

behavior, and so on, thereby completing the circle of mutually influential events from genes to environment and back again.

Another fact about genes that has not yet made its way into the psychological literature is that genetic activity does not by itself produce finished traits such as blue eyes, arms, legs, or neurons. The problem of anatomical and physiological differentiation remains unsolved, but it is unanimously recognized as requiring influences above the strictly cellular level (i.e., cell-to-cell interactions, positional influences, and so forth — Davidson, 1986; Edelman, 1988). Thus, the concept of the genetic determination of traits is truly outmoded, as is the concept of a genetically determined reaction range employed by Scarr-Salapatek in the quotation given earlier and in the various textbooks of developmental psychology cited in the introduction. (An insightful as well as witty critique of genetic determinism as applied to development is Oyama's [1985] *The Ontogeny of Information.*) The reaction-range concept is replaced by the concept of a norm of reaction, which is essentially nonpredictive because it utilizes the developmental systems view, where each new environment is expected to have a different influence on developmental outcomes that can not be stated in advance of actual empirical investigation (Platt & Sanislow, 1988). Therefore, although the genes remain an essential part of any developmental system and plasticity can not be regarded as infinite, a thoroughgoing application of the norm-of-reaction concept makes the genetic limitations on development in practice, if not in principle, unknowable (Gottlieb, 1995). Certainly, the appearance of mammalian dentition in birds (which otherwise never have teeth) under altered developmental circumstances provides striking testimony to the nonpredictability of genetic limitations on the phenotype (Kollar & Fisher, 1980).

The preceding considerations have led me to wonder about the canalizing influence of events at other levels of the developmental system, because all levels of the system may be considered potentially coequal in this respect in the developmental systems view depicted in Fig. 6.1.

POSSIBLE EXPERIENTIAL CANALIZATION OF DEVELOPMENT

In the usual interpretative framework for thinking about the canalization of development, the developing organism's interaction with its environment is viewed as the source of perturbations to the system against which the genes must buffer the developing organism in order to bring about a species-typical phenotypic outcome. The following quotation makes that point:

> Ethologists have used various models of exactly how biological regulating mechanisms control the course of development, while allowing for the

modification of development by the environment. One model, proposed by Waddington (1957), represents development as a ball rolling down an "epigenetic landscape." As the ball descends, this landscape becomes increasingly furrowed by valleys that greatly restrict the sideways movement of the ball. Slight perturbations from the developmental pathway can be corrected later through a "self-righting tendency," and the ball returns to its earlier groove. Thus, the general course of development is set, but some variation is possible because of particular environmental events. (P. H. Miller, 1989)

In the present account, I want to consider the possibility that the developing organism's usual or typical experiences can play a canalizing role that not only brings about species-specific behavior but also prevents the developing organism from being susceptible to non-species-typical forms of stimulation. Thus, the particular theoretical model that I have in mind is that normal experience helps to achieve species-specific behavioral development, and part of that process may involve making the developing organism unresponsive to extraspecific experiential influences.

EXPERIENTIAL CANALIZATION OF SPECIES-SPECIFIC PERCEPTUAL DEVELOPMENT

The concept of species-specific perception means that individuals of a given species respond in a characteristic way to certain objects; that is, they respond only to certain patterns of sensory stimulation and not to others. Usually, these patterns of stimulation are provided by other members of the species. For example, as described in chapters 3 and 5, young mallard ducklings and wood ducklings that have been hatched in incubators and never before exposed to maternal stimulation will selectively approach their own species maternal assembly call in a simultaneous auditory choice test with the mallard and wood duck maternal calls. These calls differ on critical acoustic dimensions that mallard and wood ducklings find attractive. For the mallard ducklings that feature is a call repetition rate of 4 notes/sec (±.5) and for the wood ducklings the feature is a descending frequency modulation around 1200 Hz (±200 Hz). It turns out that the ducklings' own vocalizations contain those features in an abstract way, so when the ducklings are prevented from hearing their own or sib vocalizations they do not show their usual highly selective response to the maternal call of their own species.

The fact that the ducklings have to hear their own (or sib) vocalizations to show the species-specific responsiveness to their respective maternal calls raises the possibility that exposure to such vocalizations is also playing a canalizing role in development. That is, in the absence of exposure to their own vocalizations the ducklings may become susceptible to extraspecific maternal calls, whereas exposure to their own vocalizations may render them unsusceptible to extraspecific maternal stimulation. If this hypothesis were to be supported, it would be an explicit demonstration that normally occurring *experience* is responsible for the canalization of species-specific development. It is widely believed in biology and psychology that genes or the gene-directed maturation of the sense organs and nervous system is responsible for the canalization of species-specific development in humans and animals (e.g., Fishbein, 1976; Kovach & Wilson, 1988; Lumsden & Wilson, 1980; Marler, Zoloth, & Dooling, 1981; Parker & Gibson, 1979; Scarr-Salapatek, 1976; Waddington, 1942, 1957).

In sum, the developmental systems (probabilistic epigenetic) view gives rise to the idea that canalization can take place not only at the genetic level but at all levels of the developing system (Fig. 6.1), including the developing organism's usually occurring experiences. The experiential canalization of species-specific behavioral development has only rarely been experimentally demonstrated, so the following experiment was undertaken to document that point.

EXPERIMENTAL DEMONSTRATION OF THE EXPERIENTIAL CANALIZATION OF BEHAVIORAL DEVELOPMENT

If the experiential canalization hypothesis is correct, embryonic devocalization should make the mallard embryo and hatchling more susceptible to exposure to an extraspecific maternal call, assuming the embryonic contact vocalization is acting as a canalizing or buffering experience. With the necessary control groups, five experimental conditions are necessary to test the canalization hypothesis, as follows.

In Table 6.1, X indicates the predicted preference for a particular maternal call and X---X indicates prediction of no preference. The two most critical groups are 4 and 5. It is essential to demonstrate the induction of a preference for the chicken call in devocal-isolates and the buffering effect when vocal-isolates are exposed to the embryo contact call and the chicken call. The other groups are necessary control groups.

Description of Conditions

1. *Vocal-communal.* The vocal-communal mallards were incubated together and brooded together in individual opaque boxes after hatching in groups of 15–25. Thus, they could hear themselves and each other both before and after hatching, but they could not see or physically interact with each other after hatching.

TABLE 6.1
Conditions Favoring Canalization or Malleability of Species-Specific Perceptual
Preference

| | Predicted Preference in Test | |
Condition	Mallard	Chicken
1. Vocal-communal	X	
2. Vocal-isolated	X	
3. Devocal-isolated	X----------X	
4. Devocal-isolated exposed to chicken call		X
5. Vocal-isolated exposed to chicken call and mallard duck embryo contact vocalization	X	

Note. X = preference; X---X = no preference.

2. *Vocal-isolated.* Beginning of day 24 of incubation, when the embryo is first able to vocalize, the VI mallards were put in individual soundproof incubators where they were able to hear their own vocalizations but not those of sibs. These served as controls for the devocal conditions below.

3. *Devocal-isolated.* Same as vocal-isolated, except devocalized as described earlier.

4. *Devocal-isolated, exposed to chicken call.* The DIs exposed to the chicken call received that exposure in their soundproof incubator compartments, where a speaker repetitively broadcast a single burst of a chicken maternal call for 30 min/hr from day 24 until 48 hr after hatching, at which time the birds were tested. (The call was actually on 22.5 min of the 30-min period.)

5a. *Vocal-isolated, exposed to chicken call.* These birds were treated the same as the devocal-isolated with the exception that they could hear themselves vocalize as well as being exposed to the 30 min of the chicken call each hour. (In Group 5a and in 5b described next, the embryos' heads were extricated from the shell in the same manner as in Group 3 and 4 birds, but they were not devocalized.) The prediction here, as indicated in Table 6.1, is the failure of the exposure to the chicken call to induce a preference for it because the birds have been permitted to hear their own contact call (normally occurring auditory self-stimulation).

5b. *Vocal-isolated, exposed to chicken and embryo contact calls.* These VIs were treated the same as Group 5a, except that they were exposed to 10 min of the variable-rate embryonic contact call after 30 min of exposure to the chicken call each hour. Because we did not monitor the self-produced vocalizations of the birds in Group 5a, this group provides a more precise fix on the known amount of exposure to the contact call that is required for the buffering effect. The choice of the amount of exposure to the contact call was originally arbitrary. In the 10-min period, the call was actually on for 4.6 min, which we now know from monitoring studies is at the bottom of the range produced by individual birds in isolation (4.8–6.2 min/hr, unpublished observations). This amount of "sib stimulation" is well below that which would be received under natural incubation and brooding conditions, where the birds would be exposed to as many as 8–13 conspecific siblings.

Results

As can be seen in Table 6.2, as expected, the vocal-communal and vocal-isolate birds showed a unanimous preference for the mallard call over the chicken call, whereas the unstimulated devocal-isolate did not show a preference. Further, as predicted, the devocal-isolate birds exposed to the chicken call showed a preference for the chicken call over the mallard call at 48 hr and continued to show that preference at 65 hr (retest). Also as predicted, the vocal-isolate birds exposed to the chicken call (Group 5a) did not develop a preference for it as evidenced by their behavior in the 48- and 65-hr tests. The buffering effect of the contact call is even more clearly in evidence in Group 5b, where the birds received explicit exposure to the contact call as well as the chicken call: These birds preferred the mallard over the chicken call in both tests. The results of Group 5a indicate a blockage of the chicken preference by self-stimulation but, in order to maintain the species-typical preference for the mallard call in the face of exposure to the chicken call, exposure to contact calls produced by siblings is required (Group 5b). Group 5b heard the broadcast of only one sibling's calls, which would be considerably less than would occur under normal incubation and brooding conditions in nature.

According to the chi-square test, when the preferences of the groups are

TABLE 6.2
Preferences of Mallard Ducklings in Experiential Canalization Experiment

	\multicolumn{2}{c}{*Preference*}	
Group	*Mallard Call*	*Chicken Call*
1. Vocal-communal	24***	0
2. Vocal-isolated	14***	0
3. Devocal-isolated	26	15
4. Devocal-isolated, exposed to chicken call	3	17**
Retest	2	15**
5a. Vocal-isolated, exposed to chicken call	11	15
Retest	16	12
5b. Vocal-isolated, exposed to chicken and contact calls	19	12
Retest	20*	6

*p < .01, **p < .004, ***p < .00006. (Data from Gottlieb, 1991c.)

compared across conditions, the unstimulated vocal-intact (Group 2) birds showed a greater preference for the mallard than did the unstimulated devocal-isolated (Group 3) birds. As predicted, the devocal-isolated (Group 4) birds that were exposed to the chicken call showed a greater preference for it than did the unstimulated devocal-isolated (Group 3) birds. As predicted, the stimulated devocal-isolated (Group 4) birds also showed a greater preference for the chicken call than the stimulated vocal-isolated (Group 5b) birds, which showed a greater relative preference for the mallard call. All of the above differences showed a $p < .00006$.

SIGNIFICANCE OF EXPERIENTIAL CANALIZATION OF BEHAVIORAL DEVELOPMENT

To my knowledge, this was the first explicit demonstration of the canalizing influence of normally occurring sensory experience. Exposure of mallard ducklings to their variable-rate contact call not only fosters species-specific perceptual development (i.e., ensuring selective responsiveness to the maternal call of the species), it buffers the duckling from becoming responsive to social signals from other species. In the absence of exposure to the contact call, the duckling is capable of becoming attached to the maternal call of another species even in the presence of its own species call (in simultaneous auditory choice tests). Our previous demonstration of malleability in devocalized mallard ducklings (Gottlieb, 1987b) involved the induction of a preference for either a chicken maternal call or a wood duck (*Aix sponsa*) maternal call, in which case the birds were tested with the chicken versus wood duck calls, not the mallard maternal call.

To be quite frank, I did not believe it was possible to demonstrate malleability in the presence of the species-specific maternal call. It was only when I began to think of the possible canalizing effect of experience that it became apparent that devocalization might permit such a degree of malleability that exposure to an extraspecific maternal call would override the attractiveness of the species maternal call.

Clearer thinking about what genes do and what genes don't do in individual development gave rise to the idea that canalization must take place not only at the genetic level but at all levels of the developing system, including the developing organism's usually occurring experiences (Fig. 6.1). The fact that canalizing influences are potentially present at all levels of the developing system has not been widely appreciated. Rather, as shown by the quotations in the introduction to this chapter, the widespread tendency has been to ascribe canalization exclusively to genetic activity, thereby short-circuiting developmental analysis and completely overlooking the various levels in the hierarchy of developmental systems that are necessary to produce a normal organism that exhibits species-typical behavior and psychological functioning. It is all too common to read statements such as "the vertebrate brain is fully capable of encoding stimulus information by genetic instruction" (Kovach & Wilson, 1988, p. 659)," which is merely a verbal way to close the tremendous gap between molecular biology and behavioral development. It also shows a lack of appreciation of what genes do and don't do during individual development. As it becomes more widely understood that differentiation of the nervous system (and *all* organ systems) takes place via influences above the level of the cell (Davidson, 1986; Edelman, 1987, 1988; Pritchard, 1986), a more thoroughgoing attitude or appreciation of developmental analysis will eventually supplant the verbalism of "genetic determination" and the empty metaphor of the "epigenetic landscape." I am genuinely sorry to sound so harshly critical but I do believe these ideas have provided impediments to thinking clearly about the need for conceptual and empirical analysis at all levels of the developmental systems hierarchy.

Because the particular developmental systems concept that I am advocating has only been spelled out twice before in the social or behavioral science literature (Gottlieb, 1991a, 1992), I recapitulate that view in greater detail here. The principal ideas concern the epigenetic characterization of individual development as an emergent, coactional, hierarchical system.

THE DEVELOPING INDIVIDUAL AS AN EMERGENT, COACTIONAL, HIERARCHICAL SYSTEM

The historically correct definition of epigenesis—the emergence of new structures and functions during the course of individual development—did not specify, even in a general way, how the emergent properties come into existence (Needham, 1959). Thus, there was still room for preformation-like thinking about development, which I earlier labeled the predetermined conception of epigenesis, in contrast to a probabilistic conception. (See Table 6.3 for a fuller description of probabilistic epigenesis, in addition to what is presented later.) That epigenetic development is probabilistically determined by active interactions among its constituent parts is now so well accepted that epigenesis itself is sometimes defined as the interactionist approach to the study of individual development (e.g., Dewsbury, 1978; Johnston, 1987). That is a fitting tribute to the career-long labors of Zing-Yang Kuo (1976), T. C., Schneirla (1960), and Daniel S. Lehrman (1970), the principal champions of the interaction idea in the field of psychology, and to Ashley Montagu (1977) in the field of anthropology, particularly as it applies to the study of neural, behavioral, and psychological development.

TABLE 6.3
Two Versions of Epigenetic Development

Predetermined Epigenesis:
Unidirectional Structure–Function Development

Genetic Activity (DNA → RNA → Protein) → Structural Maturation → Function,
Activity, or Experience

Probabilistic Epigenesis:
Bidirectional Structure–Function Development

Genetic Activity (DNA ↔ RNA ↔ Protein) ↔ Structural Maturation ↔ Function,
Activity, or Experience

As applied to the nervous system, structural maturation refers to neurophysiological and neuroanatomical development, principally the structure and function of
nerve cells and their synaptic interconnections. The unidirectional structure–function
(S–F) view assumes that genetic activity gives rise to structural maturation, which
then leads to function in a nonreciprocal fashion, whereas the bidirectional view
holds that there are constructive reciprocal relations between genetic activity, maturation, and function. In the unidirectional view, the activity of genes and the maturational process are pictured as relatively encapsulated or insulated so that they are
uninfluenced by feedback from the maturation process or function, whereas the bidirectional view assumes that genetic activity and maturation are affected by function,
activity, or experience. The bidirectional or probabilistic view applied to the usual
unidirectional formula calls for arrows going back to genetic activity to indicate
feedback serving as signals for the turning off and turning on of genetic activity to
manufacture protein. The usual view calls for genetic activity to be regulated by the
genetic system itself in a strictly feedforward manner. That the bidirectional view is
correct all the way to the level of DNA is evidenced by the experimental results of
researchers such as Zamenhof and van Marthens (1978), Uphouse and Bonner
(1975), Grouse et al. (1980), and Hydén and Egyházi (1962, 1964), among others. In
addition, there is now in the literature the accepted phenomenon of sensory-evoked
immediate early gene expression, as described in the text.

Note. Here and in the text I have presented the DNA → RNA → protein pathway in a
somewhat oversimplified manner, disregarding the fact that a number of crucial events
intervene between RNA and protein formation. In fact, according to Pritchard (1986),
dozens of known factors intervene between RNA activity and protein synthesis! Thus, it
is an oversimplification to imply that DNA and RNA alone produce specific proteins—
other factors (e.g., cytoplasm) contribute to the specificity of the protein.

Thus, it seems appropriate to offer a new definition of epigenesis that
includes not only the idea of the emergence of new properties but also the
idea that the emergent properties arise through reciprocal interactions
(coactions) among already existing constituents. Somewhat more formally
expressed, the new definition of epigenesis would say that *individual
development is characterized by an increase of complexity of organization—that is, the emergence of new structural and functional properties and
competencies—at all levels of analysis* (molecular, subcellular, cellular,
organismic) *as a consequence of horizontal and vertical coactions among its
parts, including organism-environment coactions.* Horizontal coactions are

those that occur at the same level (gene–gene, cell–cell, tissue–tissue, organism–organism), whereas vertical coactions occur at different levels (gene–cytoplasm, cell–tissue, behavioral activity–nervous system) and are reciprocal, meaning that they can influence each other in either direction, from lower to higher, or from higher to lower, levels of the developing system. For example, the sensory experience of a developing organism affects the differentiation of its nerve cells, such that the more experience the more differentiation and the less experience the less differentiation. (For example, enhanced activity or experience during individual development, whether spontaneous or evoked, causes more elaborate branching of dendrites and more synaptic contacts among nerve cells in the brain [Corner, 1994; Greenough & Juraska, 1979; Shatz, 1990, among many others].) Reciprocally, the more highly differentiated nervous system permits a greater degree of behavioral competency and the less differentiated nervous system permits a lesser degree of behavioral competency. Thus, the essence of the probabilistic conception of epigenesis is the bidirectionality of structure–function relationships, as depicted in Fig. 6.1 and elaborated in Table 6.3.

DEVELOPMENTAL CAUSALITY (COACTION)

Behavioral (or organic or neural) outcomes of development are a consequence of *at least* (at minimum) *two* specific components of coaction (e.g., person–person, organism–organism, organism–environment, cell–cell, nucleus–cytoplasm, sensory stimulation–sensory system, activity–motor behavior). The cause of development — what makes development happen — is the relationship of the two components, not the components themselves. Genes in themselves cannot cause development any more than stimulation in itself can cause development. Even at the genetic level itself, it is coaction among genes that mutually influence each other that plays an indispensable role in development. It is possible, for example, that it is rearrangements or reshuffling of already existing genes that brings about evolution, rather than the addition of new genes (review beginning on p. 293 in John & Miklos, 1988). This is an idea that was first put forward in a very tentative way by William Bateson in 1914 (cited in Gottlieb, 1992, pp. 82–83). So, the concept of causality as interaction or coaction applies to the level of genes as well as the neural and behavioral levels of analysis (as well as between levels of analysis). It is known, for example, that fruit flies, mice, chimpanzees, and humans share a number of the same genes (reviews in John & Miklos, 1988; Wahlsten & Gottlieb, 1996), despite their great differences in appearance, neural and behavioral organization, and psychological functioning.

When we speak of coaction as being at the heart of developmental analysis or causality, what we mean is that we need to specify some relationship between at least two components of the developmental system. The concept used most frequently to designate coactions at the organismic level of functioning is *experience*; experience is thus a relational term. As documented earlier in chapter 4, experience can play at least three different roles in anatomical, neurophysiological, and behavioral development. It can be necessary to sustain already induced states of affairs (*maintenance function*), it can temporally regulate when a feature appears during development (*facilitative function*), and it is necessary to bring about a state of affairs that would not appear unless the experience occurred (*inductive function*). Because social and behavioral scientists focus on postnatal development, it is important to recall that it is inductive interactions during embryonic or prenatal development that cause cellular differentiation (Hamburger, 1988; Wessells, 1977).[2]

Because developing systems are by definition always changing in some way, statements of developmental causality must also include a *temporal* dimension describing when the experience or organic coactions occurred. For example, one of the earliest findings of experimental embryology had to do with the differences in outcome according to the time during early development when tissue was transplanted. When tissue from the head region of the embryo was transplanted to the embryo's back, if the transplantation occurred early in development the tissue differentiated according to its new surround (i.e., it differentiated into back tissue), whereas if the transplant occurred later in development the tissue differentiated according to its previous surround so that, for example, a third eye might appear on the back of the embryo. These transplantation experiments not only demonstrated the import of time but also showed the essentially coactional nature of embryonic development.

SIGNIFICANCE OF COACTION FOR INDIVIDUAL DEVELOPMENT

The early formulation by August Weismann (1894) of the role of the hereditary material (what came to be called genes) in individual development

---

[2]A reader with philosophical interests may note that this paragraph describes causality as interaction or coaction, which is a significant departure from the Aristotelean concept of self-actional causes that are still used in science. In the same year (1991) that my suggestion was made, a philosopher of science analyzing the constructs of physics also came on the notion that "causation is interaction" (Müller, 1991, p. 170). I return to this new conception of causality in chapter 8. Meanwhile, it is gratifying to see others moving in this same direction: "The basic error of these entity based accounts is that they take networks of co-defining, co-contructing causes and attribute control to just one element in the network" (Gray, 1992, p. 194).

incorrectly held that different parts of the genome or genic system caused the differentiation of the different parts of the developing organism, so there were thought to be genes for eyes, genes for legs, genes for toes, and so forth. Hans Driesch's experiment (1908/1929), in which he separated the first two cells of a sea urchin's development and obtained a fully formed sea urchin from each of the cells, showed that each cell contained a complete complement of genes. This means that each cell is capable of developing into any part of the body, a competency that was called *equipotentiality* or *pluripotency* in the jargon of the early history of experimental embryology and *totipotency* and *multipotentiality* in today's terms (e.g., DiBerardino, 1988).

Each cell does not develop into just any part of the body, even though it has the capability of doing so. Each cell develops in accordance with its surround, so cells at the anterior pole of the embryo develop into parts of the head, cells at the posterior pole develop into parts of the tail end of the body, cells in the foremost lateral region of the embryo develop into forelimbs, and those in the hindmost lateral region develop into hindlimbs, the dorsal area of the embryo develops into the back, and so on. Although we do not know what actually causes cells to differentiate appropriately according to their surround, we do know that it is the cell's interaction with its surround, including other cells in that same area, that causes the cell to differentiate appropriately. The actual role of genes (DNA) is not to produce an arm or a leg or fingers, but to produce protein (through the coactions inherent in the formula DNA → RNA → protein in Table 6.3). The specific proteins produced by the DNA–RNA–cytoplasm coaction are influenced by coactions *above the level of DNA–RNA coaction.*

Some readers might not have known that when cells are first developed or born they are capable of becoming a cell in any organ system of the body—they are equipotential. It is only gradually during the course of further development that a cell becomes a nerve cell or a liver cell and so on by virtue of coactions with its neighbors. Even when a cell is beginning to become a nerve cell, there continue to be coactional options on precisely what part of the brain it will become (Barbe, 1996). This universal phenomenon of development cannot be understood without the concept of causality described earlier (causality as coaction or interaction).

In sum, when certain scientists refer to behavior or any other aspect of organismic structure or function as being "genetically determined," they are not mindful of the fact that genes synthesize protein (not behavior) and that they do so in the context of a developmental system of higher influences. Thus, for example, as experiments on the early development of the nervous system have demonstrated, the amount of protein synthesis itself is regulated by neural activity, once again demonstrating the bidirectionality and coaction of influences during individual development (e.g., Born & Rubel, 1988; summaries in Changeux & Konishi, 1987).

THE HIERARCHICAL SYSTEMS VIEW

Much has been written about the holistic or systems nature of individual development, beginning as early as Driesch (1908/1929) and Smuts (1926). In fact, there is no other way to envisage the manner in which development must occur if a harmoniously functioning, fully integrated organism is to be its product.

So far we have dealt with the concepts of emergence and coaction as they pertain to the development of individuals. The notion of hierarchy, as it applies to individual development, simply means that coactions occur vertically as well as horizontally in all developmental systems. All the parts of the system are capable of influencing all the other parts of the system, however indirectly that influence may manifest itself. Consonant with Sewall Wright's (1968) and Paul Weiss's (1959) depiction of the developmental system, the organismic hierarchy proceeds from the lowest level, that of the genome or DNA in the nucleus, to the nucleus in the cytoplasm in the cell, to the cell in a tissue, to the tissue in an organ, the organ in an organ system, the organ system in an organism, the organism in an environment of other organisms and physical features, the environment in an ecosystem, and so on back down through the hierarchical developmental system (review by Grene, 1987; Salthe, 1985).

A dramatic developmental effect traversing the many levels from the environment back to the cytoplasm of the cell is shown by the experiments of Victor Jollos in the 1930s and Mae-Wan Ho in the 1980s. In Ho's experiment (1984), an extraorganismic environmental event such as a brief period of exposure to ether occurring at a particular time in embryonic development can alter the cytoplasm of the cell in such a way that the protein produced by DNA–RNA–cytoplasm coaction eventually becomes a second set of wings (an abnormal "bithorax" condition) in place of the halteres (balancing organs) on the body of an otherwise normal fruit fly. Obviously, it is very likely that "signals" have been altered at various levels of the developmental hierarchy to achieve such an outcome. (Excellent texts that describe the many different kinds of coactions that are a necessary and normal part of embryonic development are N. K. Wessells' [1977] *Tissue Interactions and Development* and, more recently, for the nervous system, Gerald Edelman's [1987] *Neural Darwinism.*)

It happens that when the cytoplasm of the cell is altered, as in the experiments of Jollos and Ho, the effect is transgenerational such that the untreated daughters of the treated mothers continue for a number of generations to produce bithorax offspring and do so even when mated with males from untreated lines. Such a result has evolutionary as well as developmental significance, which, to this date, have been little exploited because the neo-Darwinian, modern synthesis does not yet have a role in evolution for anything but changes in genes and gene frequencies in evolution: Epigenetic development above the level of the genes has not yet been incorporated into the modern synthesis (Futuyma, 1988; Gilbert, Opitz, & Roff, 1996; Løvtrup, 1987).

Another remarkable organism–environment coaction occurs routinely in coral reef fish. These fish live in spatially well-defined, social groups in which there are many females and few males. When a male dies or is otherwise removed from the group, one of the females initiates a sex reversal over a period of about 2 days in which she develops the coloration, behavior, and gonadal physiology and anatomy of a fully functioning male (Shapiro, 1981). Such sex reversals keep the sex ratios about the same in social groups of coral reef fish. Apparently, it is the higher ranking females that are the first to change their sex, and that inhibits sex reversal in lower ranking females in the group. Sex reversal in coral reef fish provides an excellent example of the vertical dimension of developmental causality.

The completely reciprocal or bidirectional nature of the vertical or hierarchical organization of individual development is nowhere more apparent than the responsiveness of cellular or nuclear DNA itself to behaviorally mediated events originating in the external environment of the organism (e.g., the early, undercited experiments of Hydén & Egyházi, 1962, 1964). The major theoretical point is that the genes are part of the developmental system in the same sense as other components (cell, tissue, organism), so genes must be susceptible to influence from other levels during the process of individual development. DNA produces protein, cells are composed of protein, so, *within any species* there must be a high correlation between the *size* of cells, amount of protein, and quantity of DNA, and there must also be a high correlation between the *number* of cells, amount of protein, and quantity of DNA, and so there is (Cavalier-Smith, 1985; Mirsky & Ris, 1951).[3] For our developmental-behavioral/psychological purposes, it is most interesting to focus on the reaction of the developing brain to an organism's experience, and we find that the amount of protein in the developing rodent and chick brain is influenced by two sorts of environmental input: nutrition and sensorimotor experience. Undernutrition and "supernutrition" produce newborn rats and

---

[3]Despite these relations between quantity of DNA and some phenotypic feature in individual organisms, there is no evolutionary relationship between quantity of DNA (genome size) and organismic "complexity" (Mirsky & Ris, 1951; Sparrow, Price, & Underbrink, 1972). Furthermore, there is no relationship between the total estimated number of genes coding for protein (structural genes) and the estimated number of neurons in the brains of animals ranging from fruit flies (~250,000 neurons) to humans (~85 billion neurons) (Miklos & Edelman, 1996). The well-studied nematode *Caenorhabdhitis elegans* has approximately 14,000 structural genes and 302 neurons, whereas the fruit fly *Drosophila melanogaster* has about 12,000 structural genes and roughly 250,000 neurons. I review this topic further in chapter 9.

chicks with lower and higher quantities of cerebral protein, respectively (Zamenhof & van Marthens, 1978, 1979). Similar cerebral consequences are produced by extreme variations (social isolation, environmental enrichment) in sensorimotor experience during the postnatal period (Renner & Rosenzweig, 1987).

Because the route from DNA to protein is through the mediation of RNA (DNA → RNA → protein), it is significant for the present theoretical viewpoint that social isolation and environmental enrichment produce alterations in the complexity (diversity) of RNA sequences in the brains of rodents. (RNA complexity or diversity refers to the total number of nucleotides of individual RNA molecules.) Environmental enrichment produces an increase in the complexity of expression of RNA sequences, whereas social isolation results in a significantly reduced degree of RNA complexity (Grouse, Schrier, Letendre, & Nelson, 1980; Uphouse & Bonner, 1975). These experientially produced alterations in RNA diversity are specific to the brain. When other organs are examined (e.g., liver), no such changes are found.

The material just reviewed indicates that sensory stimulation must be activating DNA (turning genes on) in the individual experiencing the sensory stimulation. In fact, there is now so much evidence for the phenomenon that it has a name: *immediate early gene expression* (reviews by Armstrong & Montminy, 1993; Ginty, Bading, & Greenberg, 1992; Morgan & Curran, 1991; Sheng & Greenberg, 1990). An important aspect of sensory-evoked immediate early gene expression is that it is not confined to early development but continues throughout life (e.g., Mack & Mack, 1992). A brief history and a model of how immediate early gene expression may work in the nervous system are presented in Curran and Morgan (1987). (I will review the theoretical significance of immediate early gene expression at greater length in chapter 8.)

NONLINEAR CAUSALITY

Because of the emergent nature of epigenetic development, another important feature of developmental systems is that causality is often not "linear" or straightforward. In developmental systems the coaction of X and Y often produces W rather than more of X or Y, or some variant of X or Y. Another, perhaps clearer, way to express this same idea is to say that developmental causality is often not obvious. For example, in the research described earlier, we saw that mallard duck embryos had to hear their own vocalizations prior to hatching if they were to show their usual highly specific behavioral response to the mallard maternal assembly call after hatching. If the mallard duck embryo was deprived of hearing its own or sib vocalizations, it lost its species-specific perceptual specificity. To the human ear, the embryo's vocalizations sound nothing like the maternal call. It turned out, however, that there are certain rather abstract acoustic ingredients in the embryonic vocalizations that correspond to critical acoustic features that identify the mallard hen's assembly call. In the absence of experiencing those ingredients, the mallard duckling's auditory perceptual system is not completely "tuned" to those features in the mallard hen's call, and it responds to the calls of other species that resemble the mallard in these acoustic dimensions. The intricacy of the developmental causal network revealed in these experiments proved to be striking. Not only must the duckling experience the vocalizations as an embryo (the experience is ineffective after hatching), the embryo must experience *embryonic* vocalizations. That is, the embryonic vocalizations change after hatching and no longer contain the proper ingredients to tune the embryo to the maternal call.

Prenatal nonlinear causality is also nonobvious because the information, outside of experimental laboratory contexts, is usually not available to us. For example, the rate of adult sexual development is retarded in female gerbils that were adjacent to a male fetus during gestation (Clark & Galef, 1988). To further compound the nonobvious, the daughters of late-maturing females are themselves retarded in that respect — a transgenerational effect!

In a very different example of nonobvious and nonlinear developmental causality, Cierpal and McCarty (1987) found that the so-called spontaneously hypertensive (SHR) rat strain employed as an animal model of human hypertension is made hypertensive by the pups coacting with their mothers after birth. When SHR rat pups are suckled and reared by normal rat mothers after birth they do not develop hypertension. It appears that there is a "hyperactive" component in SHR mothers' maternal behavior that causes SHR pups to develop hypertension (Myers, Brunelli, Shair, Squire, & Hofer, 1989; Myers, Brunelli, Squire, Shindeldecker, & Hofer, 1989). The highly specific coactional nature of the development of hypertension in SHR rats is shown by the fact that normotensive rats do not develop hypertension when they are suckled and reared by SHR mothers. Thus, although SHR rat pups differ in some way from normal rat pups, the development of hypertension in them nonetheless requires an interaction with their mother; it is not an inevitable outcome of the fact that they are genetically, physiologically, and/or anatomically different from normal rat pups. This is good example of the *relational* aspect of the definition of experience and developmental causality offered earlier. The cause of the hypertension in the SHR rat strain is not in the SHR rat pups or in the SHR mothers but in the nursing relationship between the SHR rat pups and their mother.

Another example of a nonlinear and nonobvious developmental experience undergirding species-typical behavioral development is Wallman's (1979) demonstration that if chicks are not permitted to see their toes during the first 2 days after hatching, they do not eat or pick up mealworms as chicks normally do. Instead, the chicks stare at the mealworms. Wallman suggested that many features of the usual rearing environment of infants may offer experiences that are necessary for the expression of species-typical behavior. A further example of nonlinear and nonobvious developmental experience is Masataka's (1994) finding that laboratory-born squirrel monkeys develop a species-typical fear of snakes only if they have been fed live insects as part of their diet. Laboratory-born monkeys fed only fruit or monkey chow do not develop the species-typical fear of snakes. Yet another example of this phenomenon is Cramer, Pfister, and Haig's (1988) finding that the normal preweaning experience of nipple shifting during nursing in rat pups significantly influenced their later success in solving a spatial maze problem (eight-arm radial maze). The early experience contributes to adopting a "win-shift" strategy that is essential to solve the maze problem efficiently. Finally, D. B. Miller (in press) recently reviewed the phenomenon of nonlinear and nonobvious developmental causality in the context of his own research program.

THE UNRESOLVED PROBLEM OF DIFFERENTIATION

The nonlinear, emergent, coactional nature of individual development is well exemplified by the phenomenon of *differentiation*, whereby a new kind of organization comes into being by the coaction of preexisting parts. If genes directly caused parts of the embryo, then there would be less of a problem in understanding differentiation. Because the route from gene to mature structure or organism is not straightforward, differentiation poses a significant intellectual puzzle. The problem of differentiation also involves our limited understanding of the role of genes in development.

It has been recognized since the time of Driesch's (1908/1929) earth-shaking experiments demonstrating the genetic equipotentiality of all cells of the organism that the chief problem of understanding development was that of understanding why originally equipotential cells actually do become different in the course of development; that is, how is it that they differentiate into cells that form the tissues of very different organ systems? The problem of understanding development thus became the problem of understanding cellular differentiation. We still do not understand differentiation today, and it is quite telling of the immense difficulty of the problem that today's theory of differentiation is very much like the necessarily vaguer theories put forth by E. B. Wilson in 1896 and T. H. Morgan in 1934 (reviewed in Davidson, 1986), namely, that ultimate or eventual cellular differentiation is influenced by an earlier coaction between the genetic material in the nucleus of the cell with particular regions of the cytoplasm of the cell. Some of the vagueness has been removed in recent years by the actual determination of regional differences in the cytoplasm (extensively reviewed by Davidson, 1986). Thus, the protein resulting from locale or regional differences of nucleo-cytoplasmic coaction is biochemically distinct, which, in some as yet unknown way, influences or biases its future course of development. For example, proteins with the same or similar biochemical makeups may stay together during cellular migration during early development and thus eventually come to form a certain part of the organism by the three-dimensional spatial field considerations of the embryo mentioned earlier.

Although the actual means or mechanisms by which some cells become one part of the organism and others become another part are still unresolved, we do have a name for the essential coactions that cause cells to differentiate: They are called embryonic *inductions* (review in Hamburger, 1988). The nonlinear hallmark of developmental causality is well exemplified by embryonic induction, in which one kind of cell (A) coacting with a second kind of cell (B) produces a third kind of cell(C). For example, if left in place, cells in the upper third of an early frog embryo differentiate into nerve cells; if removed from that region, those same cells can become skin cells. Equipotentiality and the critical role of spatial position in determining differentiation in the embryo are well captured in a quotation from the autobiography of Hans Spemann, the principal discoverer of the phenomenon of embryonic induction: "We are standing and walking with parts of our body which could have been used for thinking if they had been developed in another position in the embryo" (transl. B. K. Hall, 1988, p. 174). It might have been even more striking—and equally correct—if Spemann had elected to say, "We are sitting with parts of our body which could have been used for thinking . . ."!

Even if we don't yet have a complete understanding of differentiation, the facts at our disposal show us that epigenesis is correctly characterized as an emergent, coactional, hierarchical system that results in increasingly complex biological, behavioral, and psychological organization during the course of individual development.

SUMMARY AND CONCLUSIONS

In sum, the genes are part of the developmental system and are not inviolate or immune to influences from other levels of the system as one sometimes reads in the biological literature from August Weismann to the present day.

For example, the eminent evolutionary biologist Ernst Mayr has written, "the DNA of the genotype does not itself enter into the developmental pathway but simply serves as a set of instructions" (Mayr, 1982, p. 824). Rather to the contrary, as has been demonstrated repeatedly since Hydén and Egyházi's behavioral research in the early 1960s, individual experience alters gene expression during ontogenetic development. It would seem of great importance for developmental psychologists and other social and behavioral scientists to be fully aware of this momentous change in our knowledge of genetic activity during individual development, along with the fact that genes do not by themselves produce differentiated phenotypic traits. One can hope that the immense gap between molecular biology and developmental psychology will one day be filled with facts as well as valid concepts.

7

Modifiability of Species-Specific Behavior

In an earlier chapter, I listed the usual defining features of instinctive behavior as species-typical or species-specific, unlearned (not a consequence of conditioning [S-S or S-R associative learning]), adaptive (survival value), responsive to a narrow class of sensory configurations ("sign stimuli" or "releasers" of classical ethology) in advance of exposure to these specific configurations (usually aspects of the external anatomy or behavior of other members of the species as, e.g., the maternal call). I left off one feature that the older definitions would have included and a few workers might still include when their faculties are failing them: unmodifiability. Zing-Yang Kuo, more than any other single scientist, showed the modifiability of various otherwise instinctive behaviors in dogs, cats, fish, and other species, so that defining feature is no longer a fixture, at least for most animal behaviorists and ethologists. Kuo's procedure (reviewed in 1976) was to substantially alter the usual developmental experience of his animals, thereby demonstrating that such animals do not evince species-typical behavior in the face of an altered experiential background or developmental history. The conclusion to be drawn is that instinctive behavior is not a direct or predetermined consequence of having a certain genotype, a fact that is now rather widely appreciated among animal behaviorists, if not among all students of human behavior.

COMPARISON OF MALLEABILITY IN MALLARD AND WOOD DUCKLINGS

The question here is whether the species-specific auditory preference of wood ducklings for the maternal assembly of their species is more modifiable than that of mallard ducklings for their species' maternal call. The measure of malleability employed here is whether a preference for the other species' maternal assembly call can be induced by embryonic and postnatal exposure to the call.

The possibility that the species-specific auditory perceptual development of wood ducklings might be more malleable than that of mallards stemmed from the earlier finding of a difference in the effectiveness of auditory self-stimulation for auditory species identification in the two species. Although a very small amount of solely embryonic auditory self-stimulation serves to maintain the mallard duckling's selective response to the mallard maternal call (delayed devocalization experiment in chapter 3), the wood duckling's development of perceptual specificity for the critical acoustic feature of the wood duck maternal call requires exposure to sibling vocalizations. In other words, the wood duckling must be exposed to sibling vocalizations despite the fact that the aurally isolated wood duckling vocalizes copiously and thus provides itself with considerable auditory self-stimulation (chapter 5).

Whether the greater dependence on environmentally produced auditory stimulation for species identification in the wood duckling signifies a developmentally more plastic perceptual system than in the mallard is a question of considerable interest, as it could provide some insight into experiential mechanisms relating to constraints on perceptual development. It would seem plausible that, to the extent that an organism's species-specific perceptual development is dependent on environmentally or socially produced stimulation, such a perceptual system may be more open to extraneous or alien forms of stimulation, in contrast to an organism that effectively generates its own stimulation for species-specific perceptual development. This conjecture was the hypothesis under test in the present experiment.

To pursue the question of the relative malleability of species-specific auditory perceptual development in mallard and wood ducklings, communally reared ducklings of each species were exposed to a recording of the other species' maternal call from 2 days before hatching until 2 days after hatching, at which time they were given a simultaneous auditory choice test with the mallard and wood duck maternal calls. In order to control for the amount of exposure to sibling calls, which could be especially critical to "protecting" the wood ducklings from the influence of the mallard call, each species also received exposure to recorded sibling calls that were known to be effective in promoting the specificity of response to their own species' maternal call (the distress call in the case of wood ducklings and the embryonic contact call in the mallards). As can be seen in Fig. 7.1, the mallard and wood duck maternal calls are acoustically quite different.

As shown in Table 7.1, each species had a preference for its own species

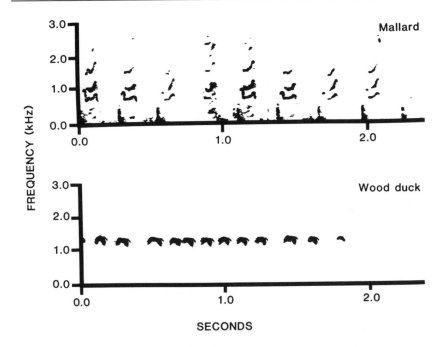

FIG. 7.1. Sonagrams of mallard and wood duck maternal calls.

maternal call without any prior exposure to the call. When the mallard and wood ducklings were given embryonic and postnatal exposure to the other species' maternal call, the mallards' preference for their own species' call was unaffected, whereas the wood ducklings showed a strong preference for the mallard maternal call over the wood duck maternal call.

The results support the hypothesis that species that rely on environmental or social experience in developing their species-specific perception are apt to be more malleable with respect to species-specific perceptual development than are animals that can promote their own species-typical or normal development by self-stimulation. Although this hypothesis seems reasonable and plausible, it will of course be necessary to see it supported under other conditions and with other species before its generality can be known. Possibly the main conceptual contribution of the present research is to call attention to how little is known about the ontogenetic, ecological, or evolutionary conditions that favor malleability or its counterpart, canalization.

Why a species that is dependent on sibling (social or environmental) stimulation should be more readily modifiable than one in which self-stimulation is effective is a question for which there is as yet no answer. The only vague sort of analogy that comes to mind with respect to the

TABLE 7.1
Preference of Mallard and Wood Ducklings in Simultaneous Auditory Choice Tests with
Mallard and Wood Duck Maternal Calls

	Preference	
Auditory Experience	Mallard Call	Wood Duck Call
Mallard ducklings		
No extrasibling		
auditory stimulation	23***	2
Stimulated with wood		
duck maternal call and mallard embryo		
variable-rate contact calls	19**	3
Wood ducklings		
Stimulated with wood		
duckling distress calls	1	13*
Stimulated with		
mallard maternal call and wood duckling		
distress calls	24***	3

*p = .003, **p = .001, ***p = .0001. (Data from Gottlieb, 1987c.)

correlation of environmental susceptibility with malleability and self-stimulation with canalized development is the bird-song-learning literature. All avian species that sing must be exposed to an environmental model, with self-stimulation being insufficient to develop the species-typical song repertoire (reviews in Kroodsma & Miller, 1982). In contrast, in birds that utter calls rather than song (e.g., chickens, doves, ducklings), deafening seems to have little effect, so the calling repertoire (species-specific motor behavior) is believed to develop more or less normally without auditory experience (Konishi, 1963; Nottebohm & Nottebohm, 1971).[1] However, there is some evidence that self-stimulation does play a critical role in certain aspects of early perceptual development. In their now classic experiments, Held and Hein (1963) showed that self-produced locomotion is essential for the development of visually guided behavior in kittens. Bertenthal, Campos, and Barrett (1984) extended those findings by showing how self-produced

---

[1]I say "believed to develop more or less normally without auditory experience," because the main technique for the analysis of calling (and singing) in deafened birds has been the Kay Sonagram, which does not reveal the fine details of the vocalizations. When finer, computer-augmented analyses are made of the calls (Scoville, 1982) and songs (Kroodsma & Konishi, 1991) of deafened ducklings and eastern phoebes, respectively, there are features of the vocalizations that depart from normal. (I thank Evan Balaban, 28 February 1996, for sharing his as yet unpublished acoustical analyses of the deafened phoebes in the Kroodsma and Konishi study. Scoville's unpublished analysis of the calls of deafened ducklings is contained in his doctoral dissertation [1982].)

locomotion facilitates emotional, cognitive, and social development in human infants. Thus, although there now are several demonstrations of the importance of self-stimulation in early development, there is not yet any analytic information on *why* self-stimulation is so efficacious, nor for the present purpose is the possible link between self-stimulation and canalization understood.

What the present results do show is that the wood duckling's instinctive attraction to the maternal call of its species is readily modifiable and, as we show later, that under different conditions of rearing, the vocal mallard duckling's attraction to its maternal call is also modifiable.

MODIFIABILITY IN VOCAL MALLARD DUCKLINGS

Several years after doing the previous experiment I accidentally discovered a degree of modifiability in mallard ducklings that was totally unpredictable as well as entirely unexpected. The interesting aspect of this accidental discovery is that it stemmed from rearing the hatchlings normally (i.e., in social groups) rather than in individual isolation as all of us had been doing in our studies of imprinting since the 1950s. (In the previous experiment in this chapter, the birds were reared in visual and tactile but not auditory isolation, the usual practice in studies of this kind. In other words, although they could hear each other, they could not see or make physical contact with each other, because they were housed in individual opaque cartons.) When one raises newly hatched ducklings in social groups, as would be the case in nature, they are even more imprintable (in the visual as well as the auditory modality) than when they are reared in isolation!

The fascinating aspect of these experiments is that the malleability (preference for the chicken maternal call over the mallard maternal call) did not require devocalizing the mallard duck embryos and hatchlings — being socially reared with peers in the presence of broadcasts of the chicken maternal call led to a preference for the chicken call in vocal mallard ducklings. Social rearing with peers establishes such a highly malleable state in the embryos and hatchlings that a species-atypical preference could be induced, one that overrides the usual canalizing influence of exposure to their own contact calls.

I first present the basic results and then describe our efforts to determine the organism–organism coactions bearing on the psychological mechanism that brings about the malleable state in socially reared ducklings.

EXPERIMENT 1: SOCIAL INDUCTION OF MALLEABILITY

In this study we exposed mallard duck embryos to a recording of a chicken maternal call for 30 min each hour from early on day 24 to 2 days after hatching, at which time they were given a simultaneous auditory choice test with the mallard and chicken maternal calls. Although we extracted the embryos' heads from the egg, it turned out this was unnecessary—socially reared ducklings from intact eggs also showed a preference for the chicken call. To see if the ducklings retained their preference for the chicken call, we retested the birds at 65 hr after hatching, without any further exposure to the chicken call between 48 and 65 hr.

The social rearing groups were composed of 15 ducklings each. Fig. 7.2 shows such a group prior to hatching. The socially isolated birds were placed in individual soundproof incubator compartments on day 24, where they remained until testing was completed. The isolated birds were exposed to the chicken call on the same schedule as the socially reared ducklings.

As shown in Table 7.2, the socially reared vocal mallard ducklings developed a strong and persistent preference for the chicken call over the mallard call, whereas the socially isolated birds did not develop a preference for the chicken call or show their usual preference for the mallard call.

EXPERIMENT 2: REPLICATION WITH INTACT EGGS AND BROADCAST OF CONTACT CALLS

Given that the embryos and ducklings had intact voices and were heard to vocalize, it came as rather a surprise that the social experience so completely

FIG. 7.2. Socially reared embryos in hatching tray in incubator (heads extricated from shell).

TABLE 7.2
Malleability of Socially Reared and Socially Isolated Ducklings Exposed to Chicken Call
Before and After Hatching

Rearing Condition and Auditory Experience	Test and Retest (hr)	Preference	
		Mallard Call	Chicken Call
Socially reared; stimulated with chicken maternal call	48	1	28**
	65	1	13*
Socially isolated; stimulated with chicken maternal call	48	10	16
	65	16	12

*p = .003, **p < .00006. (Data from Gottlieb, 1991c.)

Note. Here and in following tables birds were stimulated only to first test. In some cases, not all birds were retested. Here and in following tables, the standard size of each socially reared group was 15. However, the number of birds tested is most often not a multiple of 15, because not every bird in each social rearing group was tested.

overrode the previously documented canalizing influence of the duck embryo contact call. Because the result was truly unexpected (the socially reared group was planned as a control for other experiments), it seemed advisable to repeat the experiment with certain variations.

In Experiment 2, the socially reared embryos' heads were not extracted from the shell, thereby diluting but not completely obviating the embryonic social experience thought to be critical to the outcome (in devocalized, socially isolated birds the malleable period was prenatal; Gottlieb, 1987a). Second, in the present experiment, the embryos and hatchlings were exposed to the duck embryo contact call, which, in socially isolated birds, canalizes or buffers species-specific auditory perceptual development against influences such as exposure to the chicken maternal call (chapter 6). Although the birds in Experiment 1 had uttered the contact call, the present experiment determines more precisely the amount of exposure to that call. To investigate the canalizing influence of explicit exposure to the contact call, a second socially reared group was stimulated only with the chicken call. This group could also be directly compared with the social group in Experiment 1 to determine whether the extrication of the embryos' heads was an important factor. To investigate further the possible influence of extricating the embryos' heads, birds in a second socially isolated group had their heads extricated to compare with the intact socially isolated group. Finally, as a control to show that social rearing itself does not somehow alter the mallard duckling's preference for the mallard maternal call, a group of socially reared birds exposed solely to the duck embryo contact call was included.

As can be seen in Group 1 in Table 7.3, social rearing and exposure solely to the contact call did not alter the usual preference for the mallard maternal call over the chicken maternal call. Second, the socially reared birds exposed to the chicken call without having their heads extricated from the shell continued to show as strong a preference for the chicken call as the "extricated" birds in Experiment 1. Third, the socially reared birds stimulated with the chicken call and the contact call still showed a preference for the chicken call. The socially isolated birds stimulated with the chicken call in the intact condition (Group 4) showed the same pattern of performance as the isolated birds in Experiment 1 that had had their heads extricated: no preference for either call. Finally, in Group 5, extricating the embryos'

TABLE 7.3

Malleability of Socially Reared and Socially Isolated Ducklings Exposed to Chicken Maternal Call and Duck Embryo Contact Call Before and After Hatching

| | | | Preference | |
Group	Rearing Condition and Auditory Experience	Test and Retest (hr)	Mallard Maternal Call	Chicken Maternal Call
1.	Socially reared; egg intact; stimulated with duck embryo contact call	48	23****	0
2.	Socially reared; egg intact; stimulated with chicken call	48 65	2 1	23**** 12**
3.	Socially reared; egg intact; stimulated with chicken and contact calls	48	12	34***
4.	Socially isolated; egg intact; stimulated with chicken call	48 65	13 11	15 18
5.	Socially isolated; embryos' heads extricated from shell; stimulated with chicken and contact calls	48 65	19 20*	12 6

*$p = .01$, **$p = .006$, ***$p = .002$, ****$p = .00006$. Data from Gottlieb (1991c).

heads in the isolation condition and exposing them to the contact call as well as the chicken call made the contact call more effective so that the birds showed their usual preference for the mallard call, albeit a weakened preference in the first test.

EXPERIMENT 3: EMBRYONIC VERSUS POSTNATAL EXPOSURE

To determine more precisely the importance of embryonic as opposed to postnatal exposure to the chicken call in effecting a preference for it in the socially reared birds, in Experiment 3 we repeated Experiment 2 with one socially reared group receiving exposure to the chicken call and the contact call only prior to hatching, and the other socially reared group receiving the same stimulation only after hatching. In one prenatally stimulated group the eggs were intact (Group 1), whereas in a second prenatally stimulated Group (2) the embryos' heads were extricated from the shell. To keep the interval between the end of stimulation and the first test the same in each group, the prenatally stimulated birds were stimulated up until the time of the first test (4–8 hr after hatching), as were the birds stimulated postnatally (48 hr after hatching).

It is clear from the results shown in Table 7.4 that the birds were not malleable when stimulated with the chicken call and the duck embryo contact call either before or after hatching. Although the exposure did do away with the usual preference for the mallard call in the birds stimulated postnatally (no preference for chicken or mallard call), the birds stimulated prenatally were less affected by exposure to the chicken call, as shown by their preference for the mallard call in Group 2 in the retest. In addition, Group 2 showed a stronger preference for the mallard call and a weaker preference for the chicken call in the retest than Group 3 ($p < .05$), indicating a somewhat greater influence of postnatal social rearing on malleability.

Thus, although the malleability induced by social rearing is a cumulative effect of social rearing both before and after hatching, the postnatal contribution would appear to be somewhat stronger than the prenatal social effect. This view is further supported by the complete lack of any difference between prenatal Groups 1 and 2, where Group 1, being intact, would have much less social exposure than Group 2, in which the embryos' heads were extricated. Even though the rest of the bodies of the embryos were inside the shell, these embryos managed somehow to clump rather closely together in the incubator tray, much as the socially reared groups clumped together in a heap after hatching.

TABLE 7.4

Malleability of Socially Reared Ducklings Exposed to Chicken Call and Duck Embryo Contact Call Before or After Hatching

			Preference	
Group	Auditory Experience	Test and Retest (hr)	Mallard Maternal Call	Chicken Maternal Call
Prenatal				
1. Stimulated with chicken and contact calls before hatching (egg intact)		4–8	15	8
		23	28*	7
2. Stimulated with chicken and contact calls before hatching (embryos' heads extracted)		4–8	15	6
		23	20*	3
Postnatal				
3. Stimulated with chicken and contact calls after hatching		48	14	13
		65	10	10

*p < .01. Data from Gottlieb (1991c).

Prior to the present experiments, it seemed that the only way to render mallard ducklings susceptible to the influence of an extraspecific maternal call would be to take their voices away. As shown in chapter 6, if ducklings reared in social isolation were muted and exposed to the chicken call, they developed a preference for the chicken call over the mallard call. If they were reared in social isolation with their voices intact, they did not develop a preference for the chicken call. Finally, if they were reared in isolation with their voices intact and also exposed to a recording of a duck embryo contact call, the exposure to the chicken call was ineffective and they retained their usual preference for the mallard call. In all cases, the stimulation with the chicken call was from early on day 24 of embryonic development to 48 hr after hatching.

In the present experiments, it is clear that social rearing overrode the canalizing (buffering) effect of the contact call. The malleable period is prenatal in devocalized birds, in which the canalizing experience (exposure to contact call) is absent. Socially isolated mute birds do not become malleable as in the present experiments; rather, the embryos are malleable until they experience the contact call, which closes out their malleability as long as they remain in social isolation.[2] The present experiments indicate that prenatal and postnatal social rearing eventually override the canalizing influence of the contact call. Socially reared birds with intact voices that were exposed to the contact call preferred the chicken call to the mallard call. The obvious question arises: What is it about social rearing that induces malleability in these young birds?

SENSORY BASIS AND PSYCHOLOGICAL MECHANISM OF MALLEABILITY

There were two important avenues of inquiry to pursue to gain an understanding of the social induction of malleability in mallard ducklings. First, we needed to determine the sensory factor(s) that are crucial for social rearing to be effective. The obvious ones are vision (seeing each other) and somesthesis (tactile contact). Once the sensory bases of the social rearing phenomenon were established, then it was important to establish whether there is something different about the vocalizations that are (or are not) produced in socially effective and ineffective situations that may somehow be mediating the malleability effect. The socially reared ducklings spend most of their time huddled together, with their eyes closed, apparently asleep. Thus, they may be uttering the contact call less than social isolates and therefore not buffering themselves against exposure to the chicken call. Another possibility is that, with their eyes closed, they experience less distracting visual stimulation (i.e., there is less intersensory competition for their limited ability to attend) and exposure to the chicken call is more effective than in socially isolated birds, which may spend less time sleeping (or with their eyes closed). The intersensory competition hypothesis can be tested along with the reduced contact-calling hypothesis. A third possibility is that there is a difference in arousal level between the isolated and the socially reared birds; this can be examined by monitoring distress-calling as well as contact-calling in the rearing conditions and in the tests.

EXPERIMENT 4: IS TACTILE CONTACT ESSENTIAL TO MALLEABILITY?

Observations of socially reared mallard ducklings show them to clump together quietly with their eyes closed as if asleep for much of the time

---

[2]In an unpublished experiment, we exposed isolated mallard duck embryos to a chicken maternal call prior to day 24 (when they begin vocalizing), and these embryos later responded to the chicken call with a change in bill-clapping, which is ordinarily reserved only for the mallard maternal call. Thus, isolated embryos are malleable prior to exposure to their own vocalizations.

(Figs. 7.2 and 7.3). Given the importance of tactile contact in the normal social development of other species (Alberts, 1978; Harlow, 1958; Igel & Calvin, 1960; Jeddi, 1970), the first question was whether tactile contact is an essential component of the malleability effect. To answer this question, it was necessary to consider tactile contact alone, without other sensory factors. To do that, ducklings were reared in one of five ways:

1. Socially in groups where they could see one another and make physical contact (Figs. 7.2 and 7.3).
2. Socially in groups where they could see one another but not make physical contact (tactile isolates) (Fig. 7.4).
3. Reared in individual isolation (control for 4).
4. Individually reared with inanimate stuffed ducklings where the individual could make physical contact and see "siblings" but not interact with live siblings (rules out live interactive component) (Fig. 7.5).
5. Socially reared with blindfolds so they could make physical contact but not see each other (Figs. 7.6 and 7.7).

The five groups and the rationale for each are depicted in Table 7.5. I did not consider auditory exposure to siblings as one of the modalities of

FIG. 7.3. Social ducklings in rearing box.

FIG. 7.4. Social tactile-isolate in clear plastic container (placed inside social rearing box shown in Fig. 7.3).

malleability because we know exposure to sibling contact calls canalizes species-specific auditory development in favor of a preference for the mallard maternal call, thus preventing malleability in socially isolated ducklings.

FIG. 7.5. Individual duckling with stuffed ducklings.

FIG. 7.6. Blindfolded social ducklings in rearing box. The box is heated by two lamps from above. Each group is under one of the lamps.

FIG. 7.7. Blindfolded duckling.

TABLE 7.5

Five Groups in Experiment 4 and Rationale for Determining Tactile Basis of Malleability

| | Rationale | |
Group	Predicted Result[a]	Significance
1. Social	Malleable	Rules in tactile contact but does not rule out live interactive component or seeing one another
2. Blindfolded social	Malleable	Rules out need for seeing one another; rules in tactile contact
3. Social tactile-isolates	Not malleable	Rules in need for tactile contact; rules out seeing one another; rules in tactile contact
4. Individual isolation	Not malleable	Supports need for tactile contact and serves as control for 5
5. Individual with stuffed ducklings	Malleable	Rules ot live interactive component; rules in simple form of tactile contact

[a]Preference for chicken maternal call over mallard maternal call in simultaneous auditory choice test signifies malleability, whereas no preference or preference for mallard call signifies nonmalleability.

As can be seen in Table 7.6, as expected, the social group was malleable, showing a strong preference for the chicken call over the mallard call at 48 hr and in the retention test at 65 hr. The blindfolded social group also showed a strong preference for the chicken call in both tests, thereby ruling out the need for the siblings to see each other and supporting the need for solely tactile contact to foster malleability. When the social group was prevented from having tactile contact (Group 3 in Table 7.6), the ducklings were not malleable and showed a statistically significant preference for the mallard call over the chicken call, thus reflecting the need for tactile contact and indicating that the siblings seeing one another did not make them malleable. As would be predicted from the tactile contact hypothesis, individuals reared in isolation (Group 4) were not malleable. They did not show a preference for either call in both tests, thereby supporting the need for tactile contact for malleability. Finally, the single individuals reared with inanimate, stuffed ducklings were malleable, showing a preference for the chicken call in both tests. This result rules out the need for a live interactive component and rules in the need for a simple form of tactile contact to foster malleability (see Fig. 7.5).

In summary, then, the behavior of all five groups in the tests consistently supported the need for tactile contact in order for the ducklings to be malleable. Seeing one another does not play any role in fostering auditory malleability; in fact, it may actually interfere with auditory malleability (discussed later). Thus, the sensory basis for the development or maintenance of malleability in ducklings is tactile contact.

TABLE 7.6

Preferences in Simultaneous Auditory Choice Test With Mallard and Chicken Maternal Calls in Experiment 4

		Preference	
Rearing Condition	Test and Retest (hr)	Mallard	Chicken
1. Social	48	2	23***
	65	1	12**
2. Blindfolded	48	4	23**
social	65	1	24***
3. Social tactile-	48	25*	11
isolates	65	16	9
4. Individual	48	13	15
isolation	65	11	18
5. Individual with	48	4	18**
stuffed ducklings	65	4	13*

$^*p \leq .05$, $^{**}p \leq .006$, $^{***}p \leq .00006$.

EXPERIMENT 5: HOW DOES TACTILE CONTACT FUNCTION TO BRING ABOUT MALLEABILITY? SEARCH FOR A PSYCHOLOGICAL MECHANISM

Determination of what tactile contact does for the young duckling to make it malleable (or to maintain its malleability) involves a number of possible psychological mechanisms, among which are functional devocalization and optimum arousal. These hypothetical mechanisms are examined in turn.

Functional Devocalization Hypothesis

Because exposure of duck embryos and ducklings to their contact call has proven to be an experiential canalizing agent, the first and most obvious possibility is that the tactile contact arising out of social rearing (paradoxically) reduces the production of the contact call, thus rendering them malleable. If this functional devocalization is correct, that would mean that the socially reared ducklings that were precluded from tactile contact (Group 3, social tactile-isolates in Experiment 4) would be uttering more contact calls than the fully socially reared ducklings (Group 1 in Experiment 4). Thus their perceptual preference would be canalized to the mallard maternal call and they would be buffered against exposure to the chicken maternal call. To test this possibility, we counted the number of contact vocalizations in ducklings reared like Group 1 and Group 3 from hatching until the first test at 48 hr after hatching.

We observed 12 ducklings in each group every 6 hr from 18 to 48 hr after hatching. As before, the chicken call came on for 30 min/hr. At the beginning of the 30-min period, we monitored the ducklings' vocalizations for 5 min before and after the chicken call came on, and again at the end of the 30-min period for 5 min before and after the chicken call went off. The ducklings' vocalizations were categorized into three classes: (a) unambiguous contact call (low-pitched, short, fast notes, Scoville & Gottlieb, 1980; Scoville, 1982), (b) unambiguous distress call (high-pitched, long, slow notes, Scoville & Gottlieb, 1980; Scoville, 1982), and (c) uncertain.

As can be seen in Table 7.7, quite contrary to expectation, the nonmalleable tactile-isolates uttered fewer contact calls than the malleable social group, and they uttered many more distress and unclassifiable vocalizations than the social group. Thus, although there is no support for the functional devocalization hypothesis regarding exposure to the canalizing contact call, the tactile-isolates were clearly more vocally aroused and active than the socially reared birds. Perhaps their relatively high level of arousal interfered with their being able to be attentive/receptive to the chicken call. I can think of three ways whereby this interference might occur: (a) intersensory

TABLE 7.7
Overall Mean (± SD) Vocalization in Socially Reared and Tactile-Isolated
Ducklings in Experiment 5

Group	Contact	Distress	Unclassified
Functional devocalization			
Social	48.2	2.3	146.6
	(52.4)	(1.0)	(199.8)
Social Tactile-isolates	14.1**	19.9*	447.3***
	(27.1)	(27.7)	(215.3)
Intersensory interference			
Blindfolded tactile-	29.0	90.7	209.1
isolates	(50.4)	(40.5)	(207.7)

p = .001, **p** = .003, ***p*** < .00006, Mann–Whitney differences between social and social tactile-isolates in Experiment 2: functional devocalization hypothesis. U-test, statistical comparisons of blindfolded tactile-isolates vocalizations are given in text under Experiment 5: Intersensory Interference Hypothesis.

interference, in which being awake and active allows too much stimulation of the visual system that interferes with auditory learning (Gottlieb, Tomlinson, & Radell, 1989) or (b) not getting sufficient sleep, which has been shown in rodents to be important in consolidating learning and, in particular, memory (Tagney, 1973; review in Bloch, 1976). A third possibility (c) may be simply that a high level of arousal interferes with learning for neither of the two aforementioned reasons (Hebb, 1955). These hypotheses are examined in turn.

Intersensory Interference Hypothesis

One way to test the intersensory interference hypothesis would be to blindfold the tactile isolates and expose them to the chicken call. If reducing their visual input would make them malleable, that would be clear support for the intersensory interference hypothesis.

Experiment 4 was repeated as before, except that this time the tactile-isolated group was blindfolded. In addition to testing the birds for their preference for the chicken call in the mallard–chicken call choice test, the vocalizations of these birds were monitored as described earlier.

As shown in Table 7.8, the blindfolded tactile-isolates were not malleable, in that they did not develop a preference for the chicken call over the mallard call. Comparison of the preferences of the previous "sighted" (social) tactile-isolates (Experiment 4, Table 7.6) with the present blindfolded tactile-isolates revealed no statistically reliable differences (chi-squared test). Consequently, there is no support for the intersensory interference hypothesis.

Table 7.7 shows the vocalizations of the blindfolded tactile-isolates. The

TABLE 7.8
Preferences of Blindfolded Tactile-Isolated Group in Mallard Versus Chicken Call Test

| Test and Retest (hr) | Preference | |
	Mallard	Chicken
48	13	12
65	12	15

blindfolded tactile-isolates tended to utter more contact calls than the social tactile-isolation in Experiment 1 (two-tailed $p = .10$) and were much less vocal as far as producing unclassified vocalizations ($p = .0006$). In comparing the vocal activity of the nonmalleable social tactile-isolates with that of the malleable social group, the former produced fewer contact notes and more distress calls than the social group ($p = .03$ and $.01$, respectively). Thus, the blindfolded tactile-isolates and the "sighted" tactile-isolates were much more activated or aroused than the socially reared birds, which, once again, raises the possibility that their high level of arousal precluded their getting sufficient sleep for the exposure to the chicken call to be effective. To test this hypothesis we compared the awake–sleep patterns of these groups in relation to exposure to the chicken call.

Sleep Interference Hypothesis

It has been demonstrated in adult rodents that paradoxical or rapid eye movement sleep (REM) plays a role in the "fixation of the memory trace" in the hours following learning by activating the brain in a special way (Bloch, 1976; Lucero, 1970). In the standard Berkeley rodent environmental enrichment paradigm, Tagney (1973) observed that rats in the enriched condition manifest significantly more slow-wave sleep time and REM sleep time than rats reared in an impoverished condition (reared in individual cages), a rearing condition similar to the tactile-isolated group in the present study. To determine whether sleep deprivation was occurring in the groups that were not malleable, we observed the incidence of behavioral sleep patterns in the social and blindfolded social groups that were malleable and compared these patterns with those of the social tactile-isolates and the blindfolded tactile isolates that were not malleable.

We observed 12 ducklings in each of the four groups every 6 hr between 18 and 48 hr after hatching. They were observed at 5 min before the 30-min period during which the chicken call was broadcast and again at the beginning of the last 5 min of the 30-min broadcast of the chicken call. The number of birds awake in each group at each observation was recorded. Birds were recorded as being awake if they had their eyes open and were sitting erectly, standing, or moving. When the ducklings are asleep, they usually clump together in a physically relaxed way if they are not tactually isolated (compare Figs. 7.3 and 7.4).

As can be seen in Table 7.9, in the malleable social group a total of only two birds (2.8%) were awake 5 min before the chicken call was broadcast on the six occasions of observation and 88.9% of the social birds were awake at the 25th minute during the broadcast of the call. This pattern conforms very well to the conditions necessary for learning: asleep when the call was off, and awake when the call was on (Bloch, 1976). This pattern, however, was not observed in the other malleable group; 39% of the blindfolded social birds were awake before the call and only 49% were awake during the call. These figures are comparable to those of the nonmalleable blindfolded tactile-isolates. Somewhat more birds were awake before and during the call in the nonmalleable, social tactile-isolated group.

Because the awake/sleep pattern in the malleable blindfolded social group differed from the malleable social group and did not differ from the nonmalleable blindfolded tactile-isolates, under the present conditions there is no support for the notion that sleep deprivation interferes with malleability.

Excessive Arousal Hypothesis

Although highly refined behavioral (in contrast to physiological) categorizations of arousal are a staple of human infant developmental psychology (Thoman, 1990), with the exception mentioned later, such is not the case in the animal developmental literature. Arousal level has been shown to be crucial when assessing the human infant's discriminative response to sound (e.g., Williams & Golenski, 1970). Recently, Gray (1990) elegantly quantified the behavioral arousal-level continuum for newly hatched chicks in relation to discriminative auditory responding. He obtained the expected

TABLE 7.9
Mean Percentage of Birds Awake on Six Occasions From 18 to 48 hr After Hatching

	Group			
	Social (Malleable)	Social Tactile-Isolates (Not Malleable)	Blindfolded Social (Malleable)	Blindfolded Tactile-Isolates (Not Malleable)
Before	2.8	47.2	38.9	33.3
During	88.9	68.0	48.6	41.6

Note. Before: observation made 5 min before chicken call broadcast. During: observation made after chicken call had been broadcast for 25 min of 30-min broadcast period.

inverted-U- or bell-shaped curve, with low and high levels of behavioral arousal being associated with poor auditory discrimination. Gray's results are especially pertinent to the present study because he used the chicks' rate of distress peeping as the measure of behavioral arousal. High rates of distress peeping were associated with poor auditory discrimination.

The distress calling level of the nonmalleable birds in the present study (tactile-isolates and blindfolded tactile-isolates) was 20–40 times that of the malleable birds (social and blindfolded social groups), as indicated in Table 7.7.

This result suggests that extremely high levels of arousal interfere with the birds' attentiveness so that exposure to the chicken call is ineffective. If that were indeed the case, there should be a correlation between an individual's arousal level and its behavior in the test situation, with nonmalleable birds showing a significantly greater level of arousal than malleable birds. With this idea in mind, we attempted to monitor the vocalizations of individually marked birds in the social and tactile-isolated groups before and during exposure to chicken call. Viewing the birds from overhead through a mirror above their rearing box, we could not monitor the vocalizations of individuals, so other means need to be worked out to monitor individuals during exposure to the chicken call. Although it is not yet possible to monitor the vocalizations of 12 voluble individuals in a group, it is possible to monitor them individually when the birds are being observed singly in the test arena, so that procedure was used to correlate an individual's arousal level with its malleability in the test (48 hr) and retest (65 hr) situation.

Twelve social and 12 social tactile-isolated ducklings were reared in groups, exposed to the chicken call, and then tested and retested as in Experiment 1. The procedure was followed four times, yielding 48 birds in each group. The distress calls of each bird were tallied in the test and retest situation. To answer the question of whether the nonmalleable ducklings are more aroused than the malleable ducklings, the distress calls of the nonmalleable birds (those that either preferred the mallard call over the chicken call or did not respond to either call) were compared with those of the malleable birds (those that preferred the chicken call over the mallard call). Two types of analysis were performed. Because the social groups would be malleable and the tactile-isolated groups would not be malleable, if arousal level as indexed by number of distress calls is indeed involved in malleability, the malleable social group would utter fewer distress calls than the nonmalleable tactile-isolated group (between-group comparison), irrespective of preference in the auditory choice test. The second, more refined analysis involved individual ducklings within each group, with the prediction that nonmalleable ducklings would utter more distress calls than malleable ducklings within each group.

TABLE 7.10
Distress Calls Uttered During Testing and Retesting in Social and
Tactile-Isolated Groups

Group	Test	Retest
Social	158	90.5
	(24–289)	(13–280)
Social tactile-isolates	272**	233.5*
	(142–338)	(75.5–299)

*p = .05, **p = .01, Mann–Whitney U-test. Medians, with interquartile ranges in parentheses.

As shown in Table 7.10, the tactile-isolated ducklings as a group uttered more distress calls than the social group in the test and retest situations. Furthermore, in Table 7.11, within each group, the nonmalleable birds uttered more distress calls than the malleable birds in both the test and retest conditions.

SUMMARY AND CONCLUSIONS

To summarize, when the ducklings were reared under a variety of conditions designed to determine the sensory basis of malleability (various conditions designated in Table 7.5), the conclusion seems firm that tactile contact is the essential sensory modality, and that such contact is effective even if provided by stuffed ducklings. Being able to see one another is not part of the social complex involved in fostering auditory malleability. Also, tactile contact overrides the canalizing influence of auditory exposure to contact calls in that the malleable, socially reared birds produced copious

TABLE 7.11
Distress Calls Uttered During Testing and Retesting by Malleable and Nonmalleable
Birds Within Social and Tactile-Isolated Groups

Group	Test		Retest	
	Malleable	Nonmalleable	Malleable	Nonmalleable
Social	41	253**	34	316***
	(3–204)	(156.5–365)	(8–126)	(180–389)
Social tactile-isolates	192	332**	125	262.5*
	(72–242)	(272–367)	(19–262)	(215–343)

*p = .01, **p = .001, ***p = .0001, Wilcoxon rank-sum test. Medians, with interquartile ranges in parentheses.

contact calls compared with the ducklings in the other nonmalleable groups. Finally, it seems clear that tactile isolation engenders such a high degree of psychological arousal that it interferes with malleability, much as excessive arousal interferes with auditory discrimination ability in chicks (Gray, 1990) and the behavioral freezing response to the maternal call in ducklings (Miller & Gottlieb, 1981).

Thus, excessive arousal not only interferes with the malleability underlying species-atypical behavior but also with species-typical behavior. Although the behavior of the ducklings in the test and retest situations indicates that excessive arousal is associated with nonmalleability (preference for the mallard over the chicken call or no preference), it remains to be documented that individual nonmalleable birds are more aroused than malleable individuals during exposure to the chicken call prior to testing. (The results of monitoring the vocal activity of the various malleable and nonmalleable groups strongly suggest that it is excessively high arousal during exposure to the chicken call that interferes with malleability. It remains only to correlate an individual's arousal level during exposure to the chicken call with its preference in the choice test.) When sensory overload disrupts the usual auditory learning ability of mallard duck embryos (Gottlieb et al., 1989), we have ascribed that interference to a disruption of the embryo's poorly developed attentional capacity. It is tempting to suggest a similar disruption of attention by excessive arousal in the present case. An attentional deficit, coupled with behavioral overactivity, is often a demonstrable feature of learning disorders in human children (Harter, 1991).

Given the interest in intersensory influences in early behavioral development (reviews in Turkewitz & Mellon, 1989; Tees, 1990), it seems worthwhile to call attention to the quite subtle intermodal effect in the present work. The presence of tactile contact afforded by normal social interaction with siblings or sibling surrogates has a positive influence on auditory learning ability, whereas tactile deprivation interferes with auditory learning in the case of an extraspecific maternal call. Tactile sensory stimulation from siblings is not essential to the learning of the species-typical visual attributes of siblings (Dyer, Lickliter, & Gottlieb, 1989), nor is it essential to the learning of species-typical maternal calls (G. Gottlieb & P. Radell, unpublished data).

Finally, at the physiological level of analysis, with the demonstrated correlation of β-endorphins and social behavior in mammals and birds (reviewed in Hinde, 1990, p. 124), it seems reasonable to hypothesize a difference in β-endorphin levels between the ducklings in the social and tactile-isolated groups, with the former perhaps showing higher levels than the latter. This would be consonant with the finding that opiate agonists decrease distress calling in socially isolated chick, whereas opiate antagonists enhance distress calling in such chicks (Panksepp, Bean, Bishop,

Vilberg, & Sahley, 1980). Of course, the probable correlation of β-endorphin levels and social grouping may very well be specific to the social situation and thus be only indirectly correlated with malleability and nonmalleability, especially if nonmalleability reflects an attentional deficit. However, if the appropriate pharmacological intervention in the tactile-isolated birds alleviated distress calling (i.e., lowered arousal level) and they were then malleable, that would not be an uninteresting result.

8

Evolution of Probabilistic Epigenesis: History and Current Status of a Developmental-Psychobiological Systems Viewpoint

In his first paper, published in 1921, Zing-Yang Kuo assaulted the concept of instinct. His attack was based on the insight that the use of instinct as an explanatory concept was harmful to a genuine understanding of behavior because it made the analysis of development superfluous and, therefore, unnecessary. For Kuo, a psychology based on instinct was an armchair psychology and, as such, it was not an investigative science. Throughout his career he remained keenly sensitive to concepts which served as facile substitutes for experimental analyses, especially those concepts that diverted attention from an investigation of developmental processes, whether these processes be anatomical, physiological, or behavioral. This is not to say that Kuo was antitheoretical—he simply made a very strong distinction between hypotheses that promote investigation and those that do not. For him, the concept of instinct fell in the latter category. There is, after all, a very significant difference in positing, say, a reproductive instinct to account for nest building in birds instead of undertaking a developmental analysis of nest-building behavior.

In his critique, Kuo acknowledged that animals perform all manner of activities without the benefit of, or opportunity for, prior learning or imitation—for him, these performances were not simply to be taken as demonstrations of the wonderful workings of instinct; rather, they posed interesting and significant problems for ontogenetic analysis. It was his lifeling conviction (Kuo, 1976) that an explanation of an animal's behavior could be derived entirely from (a) its anatomy and physiology, (b) its current environmental setting, and (c) its individual developmental history. His own research and his critical writings can be comprehended only in the light of his belief that any analysis of behavior is incomplete if it relies mainly or exclusively on only one of these factors—all three must be taken into consideration in a comprehensive account of the development of animal and human behavior.

Building on Kuo's wholly original insights, I have added five items to the developmental analysis of instinctive behavior. On the theoretical side, (a) I have made explicit the bidirectionality of influences across the various levels of functioning, and (b) have added genetic activity to the levels of analysis. On the empirical side, (c) I have shown that the normal development of brain physiology and species-specific perception is dependent on normally occurring embryonic experience, (d) have demonstrated that the canalization of instinctive behavior can be influenced at the organism-environment experiential level in addition to the genetic level, and (e) have elucidated two alternative developmental pathways to malleability in ducklings.

Over the years, I have gradually come to appreciate that the analytic framework that has evolved is a culmination of thinking that began in the early 1800s, so I now wish to put the current theoretical and experimental results in an even broader context.

HISTORY OF DEVELOPMENTAL SYSTEMS THINKING

The current definition of epigenesis holds that individual development is characterized by an increase in novelty and complexity of organization over time—that is, the sequential emergence of new structural and functional properties and competencies—at all levels of analysis as a consequence of horizontal and vertical coactions among its parts, including organism-environment coactions (chapter 6). Our present understanding of the various defining features of epigenesis has been laboriously worked out over the past 200 years. This section briefly recounts that intellectual history.

The Triumph of Epigenesis Over Preformation

The triumph of epigenesis over the concept of preformation ushered in the era of truly developmental thinking. Namely, the notion that to understand the origin of any phenotype it is necessary to study its development in the individual. This insight has been with us since at least the beginning of the 1800s, when Etienne Geoffroy Saint-Hilaire (1825) advanced his hypothesis that the originating event of evolutionary change was an anomaly of embryonic or fetal development. The origin or initiation of evolutionary change was thus seen as a change in the very early development of an atypical individual. Although not a believer in evolution (in the sense that a species could become so modified as to give rise to a new species), Karl Ernst von Baer (1828/1966) used the description of prenatal anatomical development as a basis for classifying the relationships among species: Those that shared the most developmental features were classified together, whereas those that shared the fewest features were given a remote classification. It was von Baer who noticed that vertebrate species are much more alike in their early developmental stages than in their later stages. This was such a ubiquitous observation that von Baer formulated a law to the effect that development in various vertebrate species could be universally characterized as progressing from the homogeneous to the heterogeneous. As individuals in each species reached the later stages of their development they began to differentiate more and more away from each other, so there was less and less resemblance as each species reached adulthood.

The Birth of Experimental Embryology

Although von Baer's emphasis on the importance of developmental description represented a great leap forward in understanding the question of "what," it did not come to grips with the problem of "how;" namely, he and his predecessors evinced no interest in the mechanisms or means by which each developmental stage is brought about — it simply was not a question for them. It remained for the self-designated *experimental* embryologists of the late 1800s to ask that developmental question: Wilhelm His, Wilhelm Roux, and Hans Driesch. As His (1888, p. 295) wrote in reference to von Baer's observations:

> By comparison of [the development of] different organisms, and by finding their similarities, we throw light upon their probable genealogical relations, but we give no direct explanation of their growth and formation. A direct explanation can only come from the immediate study of the different phases of individual development. Every stage of development must be looked at as the physiological consequence of some preceding stage, and ultimately as the consequence of the acts of impregnation and segmentation of the egg.

It remained for Roux, in 1888 (Roux, 1888/1974), to plunge a hot needle into one of the two existing cells after the first cleavage in a frog's egg, thereby initiating a truly *experimental* study of embryology.

The arduously reached conclusion — the one still held today — that individual development is most appropriately viewed as a hierarchically organized system began with Hans Driesch being dumbfounded by the results of his replication of Roux's experiment. Although Roux found that killing one cell and allowing the second cleavage cell to survive resulted in a half-embryo in frogs, Driesch (reviewed in 1908; Driesch, 1908/1929) found that disattaching the first two cells in a sea urchin resulted in two fully formed sea urchins, albeit diminished in size. (When the disattachment procedure was later used in amphibians, two fully formed embryos resulted as in Driesch's experiment with sea urchins [Mangold & Seidel, 1927].) Driesch came to believe there is some nonmaterial vitalistic influence (an *entelechy*) at work in the formation of the embryo, one that will forever elude our best experimental efforts, so he eventually gave up embryology in favor of the presumably more manageable problems of psychology.

Because Driesch found that a single cell could lead to the creation of a fully formed individual, he gathered, quite correctly, that each cell must have the same prospective potency, as he called it, and could in principle become any part of the body. He thought of these cells as *harmonious-equipotential systems*. For Driesch, the vitalistic feature of these harmonious-equipotential systems is their ability to reach the same outcome or endpoint by different routes, a process that he labeled *equifinality*. Thus, in the usual case, you have two attached cleavage cells giving rise to an embryo, whereas in the unusual case, you have two separated cleavage cells, each giving rise to an embryo. Although to Driesch these experimental observations provided the most elementary or "easy" proofs of vitalism, for those still laboring in the field of embryology today they continue to provide a provocative challenge for experimental resolution and discovery (see, e.g., Miklos & Edelman, 1996).

For the present purposes, it is important to note that, if each cell of the organism is a harmonious-equipotential system, then it follows that the organism itself must be such a system. Driesch's notion of equifinality — that developing organisms of the same species can reach the same endpoint via different developmental pathways — has become an axiom of embryological systems theory.[1] In a systems view of developmental psychobiology, equifinality means (a) that developing organisms that have different early or "initial" conditions can reach the same endpoint and (b) that organisms that share the same initial condition can reach the same endpoint by different routes or pathways (cf. Ford & Lerner, 1992). Both of these points have been empirically demonstrated by the behavioral research of D. B. Miller (Miller, Hicinbothom, & Blaich, 1990) and R. Lickliter (Banker & Lickliter, 1993) in birds, and by Nöel (1989) and Carlier, Robertoux, Kottler, and Degrelle (1989), among others, in mammals. The uniquely important developmental principle of equifinality is rarely explicitly invoked in theoretical views of developmental psychology, so it may seem unfamiliar

---

[1]Egon Brunswik (1952) was the first to call attention to equifinality as an important principle of psychological development in his infrequently cited monograph for the International Encyclopedia of Unified Science, *The Conceptual Framework of Psychology*.

to many readers. K. W. Fischer's (1980, p. 513) theory of skill development in infancy and early childhood is one of the rare exceptions in that it explicitly incorporates the notion of equifinality, to wit, "different individuals will follow different developmental paths in the same skill domain . . . . The developmental transformation rules predict a large number of different possible paths in any single domain."

Microgenetic studies of human development are most likely to reveal equifinality, because under these conditions the response of individuals to the same challenge is closely monitored and described for shorter or longer periods (e.g., Kuhn, 1995). In one such study, Bellugi, Wang, and Jernigan (1994) monitored the attempted solutions of Williams' syndrome and Down's syndrome children to the block design subtest on the WISC-R. Although from the ages of 10 to 18 years, the children in both groups performed equally poorly, the attempted solutions by the Down's syndrome individuals approximated in a global way the design they were trying to copy, whereas the Williams' syndrome group uniquely failed to reproduce the correct global configuration of the blocks. As shown in Fig. 8.1, the children in both groups get the same low scores but they achieve them in very different ways (by different pathways, if you will). In another example involving a study of language development in young hearing and deaf children, both groups ended up having an arbitrary system of signs to refer to events and objects, but the hearing preschool children achieved the outcome by using the language of their adult caretakers as models, whereas the deaf preschool children, born to hearing parents who did not know sign language, developed their own arbitrary set of gestures to communicate meaningfully with peers and adults (Goldin-Meadow, in press). As a final example from the literature, in lines of mice selectively bred for high and low aggression, individuals in the low line become as aggressive as the high line if they are tested four times from day 28 to day 235 of life (Fig. 8.2; Cairns, MacCombie, & Hood, 1983). Once again, the developmental pathways to the same endpoint are different.

In this context, I hope the reader who has come this far into the book will recognize the different routes to malleability in ducklings (devocalization or social rearing) as yet another example of equifinality.

In my opinion, the principle of equifinality is one of the most important ones in developmental-psychobiological theory for understanding both normal and abnormal development. With respect to the latter, it offers a rationale for psychotherapy, for example, and the persistent tendency toward mental health that Harry Stack Sullivan (1953) recognized. The phenomenon of equifinality indicates that there is more than one developmental route to psychological well-being, and there is more than one developmental route to the same psychological disorder. It is now clear, for example, that there are several different possible prenatal and postnatal

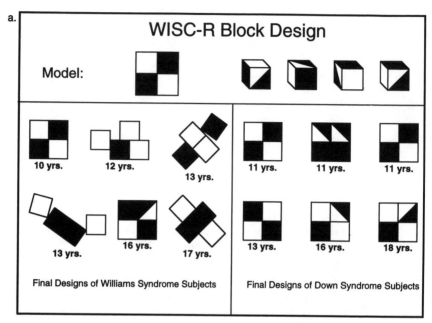

WISC-R Block Design

Model:

Final Designs of Williams Syndrome Subjects

10 yrs.   12 yrs.   13 yrs.   13 yrs.   16 yrs.   17 yrs.

Final Designs of Down Syndrome Subjects

11 yrs.   11 yrs.   11 yrs.   13 yrs.   16 yrs.   18 yrs.

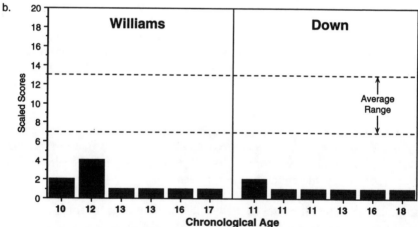

Both WMS and DNS subjects fail the WISC-R Block Design test, but in very different ways.

FIG. 8.1. Contrasting block design performance in Williams syndrome (WS) and Down's syndrome (DS). (a) Both WS and DS designs reveal striking differences in their errors. WS subjects uniquely fail to reproduce the correct global configuration of blocks. (b) However, these differences are not reflected in their quantitative scores, which are comparably low. From Bellugi, Wang, and Jernigan (1994, p. 38). Copyright © Dr. U. Bellugi, The Salk Institute, La Jolla, California.

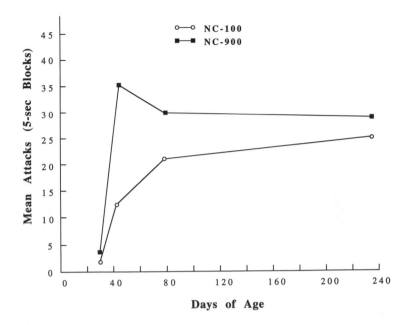

FIG. 8.2. Mean number of 5-sec blocks in which subjects of the high aggressive (NC-900) and low aggressive (NC-100) lines attacked their test partners. (The same subjects were repeatedly tested at days 28, 42, 72, and 235.) Modified from Cairns, MacCombie, and Hood (1983, p. 81).

routes to schizophrenia (reviews in Crow, 1995; Davis & Phelps, 1995), and the same will no doubt hold true for manic-depressive illness, panic, hyperactive-attention, and other such psychological disorders. To search for a single defective gene (or many nonfunctional genes) is necessary, but it is only one route to understanding the development of these psychological problems, as best exemplified when one identical twin develops a severe disorder and the other one does not, as happens with schizophrenia (Gottesman & Shields, 1982).

Systems Versus Mechanico-Reductive and Vitalistic-Constructive Viewpoints

As we move from the late 1800s to the 1930s in our overview of the precursors to our present concept of the systems nature of psychobiological development, we encounter the insights of the systems- or organismically-oriented embryologists, Paul Weiss and Ludwig von Bertalanffy, and the physiologically oriented population geneticist Sewall Wright.

In his wonderfully lucid and historically complete opus on the topic of development, *Modern Theories of Development: An Introduction to Theoretical Biology*, originally published in German in 1933, von Bertalanffy (1933/1962) introduced the system theory, as he called it, as a way of avoiding the pitfalls of machine theory, on the one hand, and vitalism, on the other. The error of the machine theory of development, as von Bertalanffy saw it, was the attempt to analyze the various aspects of the development process into their individual component parts or mechanisms, conceived of as proceeding independently of one another. Von Bertalanffy believed that the fundamental error of the classical concept of mechanism, which was adopted wholesale from physics, lay in its application of an additive point of view to the interpretation of living organisms.

> Vitalism, on the other hand, while being at one with the machine theory in analyzing the vital processes into occurrences running along their separate lines, believed these to be co-ordinated by an immaterial, transcendent entelechy. Neither of these views is justified by the facts. We believe now that the solution of this antithesis in biology is to be sought in an *organismic* or *system theory* of the organism which, on the one hand, in opposition to machine theory, sees the essence of the organism in the harmony and co-ordination of the processes among one another, but, on the other hand, does not interpret this co-ordination as vitalism does, by means of a mystical entelechy, but through the forces immanent in the living system itself. (von Bertalanffy, 1933/1962, pp. 177–178)

Nowadays, we make von Bertalanffy's point by distinguishing between theoretical and methodological reductionism. Theoretical reductionism seeks to explain the behavior of the whole organism by reference to its component parts—a derivative of the older additive, physical concept of mechanism—whereas methodological reductionism holds that not only is a description of the various hierarchically organized levels of analysis of the whole organism necessary but also that a depiction of the bidirectional traffic between levels is crucial to a developmental understanding of the individual.[2]

---

[2]As a tribute to his long and productive career in neuroembryology, the *International Journal of Developmental Neuroscience* publishes an Annual Viktor Hamburger Award Review. In 1993, the award went to Ira B. Black, who published a review on "Environmental regulation of brain trophic interactions," which detailed the influence of neural activity on multiple trophic (growth) factors during development, further attesting to the feasibility of working out the bidirectional relations depicted later in Fig. 8.5. The author (Black) himself raised that optimistic question at the conclusion of his review (p. 409): "Are we now in a position to move from environmental stimulus to impulse activity, trophic regulation, mental function and behavior . . . ?" The most recent Viktor Hamburger Award Review (1994) continues that theme with Carla Shatz's "Role for spontaneous neural activity in the patterning of connections between retina and LGN during visual system development," which is also in

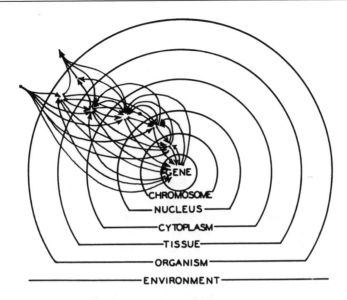

FIG. 8.3. The embryologist Paul Weiss's hierarchy of reciprocal influences moving back and forth from the lowest level of organization (gene) to the highest level (external environment). From Weiss (1959). Reprinted by permission.

For purposes of recognizing historical precedent, it is appropriate here to present the diagrams of Paul Weiss and Sewall Wright as they exemplify the strictly methodological reductionism of the hierarchically organized systems view of development. (I use the plural form of system because the various levels of organismic functioning constitute within themselves systems of analysis: the organism–environment ecological system, the nervous system, the genomic system, for example. Von Bertalanffy himself later [1950] came to use the plural form in his conception of General Systems Theory.)

In Paul Weiss's (1959) diagram of the hierarchy of reciprocal influences, as shown in Fig. 8.3, there are seven levels of analysis. The *gene* (DNA) is the ultimately reduced unit in an ever-expanding analytic pathway that moves from gene to *chromosome* — where genes can influence each other —

---

keeping with my broad definition of the term experience as spontaneous or evoked functional activity in chapter 6. Even when an organism's experience arises out of an interaction with the external environment there is an essential internal (cellular) correlate to that activity, so that is the rationale for including endogenous activity as part of the experiential process. Also, neural cells that enhance their development as a consequence of spontaneous or endogenous activity are agnostic about whether the activity arose within the nervous system or was a consequence of external stimulation. Perhaps, for some readers, it would be more appropriate to drop the term *experience* and use the term *functional activity* at all levels of analysis. To my way of thinking, as I noted earlier, *experience* and *functional activity* are synonymous.

from cell *nucleus* to cell *cytoplasm*, and from cell to tissue (organized arrangements of cells that form organ systems—the nervous system, circulatory system, musculoskeletal system, etc.), all of which make up the organism that interacts with the external environment. The entire schema represents a hierarchically organized system of increasing size, differentiation, and complexity, in which each component affects, and is affected by, all the other components, not only at its own level but at lower and higher levels as well. Thus, the arrows in Fig. 8.3 not only go upward from the gene, eventually reaching all the way to the external environment through the activities of the whole organism, but the arrows of influence return from the external environment through the various levels of the organism back to the genes.

Although the feedforward or feedupward nature of the genes has always been appreciated, the feedbackward or feeddownward influences have usually been thought to stop at the cell membrane. The newer conception is one of a totally interrelated, fully coactional system in which the activity of the genes themselves can be affected through the cytoplasm of the cell by events originating at any other level in the system, including the external environment. It is known, for example, that external environmental factors such as social interactions, changing day length, and so on can cause hormones to be secreted (review by Cheng, 1979), which, in turn, results in the activation of DNA transcription inside the nucleus of the cell (i.e., "turning genes on"). There are now many empirical examples of external sensory and internal neural events that excite and inhibit gene expression (e.g., Anokhin, Milevsnic, Shamakina, & Rose, 1991; Calamandrei & Keverne, 1994; Mauro, Wood, Krushel, Crossin, & Edelman, 1994; Rustak, Robertson, Wisden, & Hunt, 1990), thereby supporting the *bidirectionality* of influences among the various levels of analysis from gene to environment (to be discussed further).

Because Weiss was an experimental embryologist, it was probably merely an oversight that he did not explicitly include a developmental dimension in his figure. Yet another not explicitly developmental schematic of a systems view was put forward by Sewall Wright in 1968. As shown in Wright's schema (Fig. 8.4), once again, the traffic between levels is bidirectional and the activity of the genes is placed firmly inside a completely coactional system of influences. It is a small but important step to apply this way of thinking to the process of development, as I did in Fig. 6.1 in chapter 6 and reproduce here as Fig. 8.5 for the reader's convenience.

INFLUENCE OF SENSORY STIMULATION ON GENETIC ACTIVITY

As mentioned earlier, some behavioral scientists, including developmental psychologists, seem to be unaware of the fact that the genes (DNA)

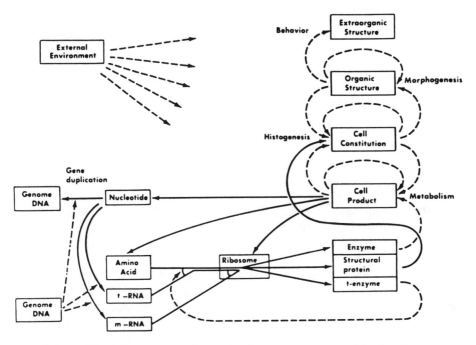

FIG. 8.4. The fully coactive or interactional system, as presented by the physiologically oriented population geneticist Sewall Wright (1968).

BIDIRECTIONAL INFLUENCES

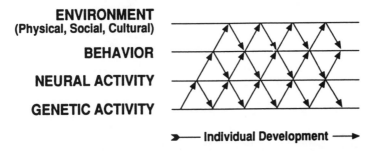

FIG. 8.5. A developmental-psychobiological systems framework. From *Individual Development and Evolution: The Genesis of Novel Behavior* by Gilbert Gottlieb. Copyright © 1991 by Oxford University Press, Inc. Reprinted by permission.

themselves are subject to influences from higher levels during the course of development, so it is useful to stress that contingency as a part of the *normal* process of development. For example, as noted in chapter 6, there is a category of genetic activity called *immediate early gene expression*, which is specifically responsive to sensory stimulation and results in a higher number of neurons in the brains of animals that have been appropriately stimulated and a deficiency in the number of cortical neurons in animals that have been deprived of such normal sensory stimulation (e.g., Rosen, McCormack, Villa-Komaroff, & Mower, 1992, and references therein). It was not so long ago that neuroscientists of very high repute, including at least one eventual Nobel prize winner, were writing in a vein that would seem to make sensory-stimulated immediate early gene expression an impossibility, much less an important feature of normal neurobehavioral development. For example, Roger Sperry wrote in 1951 (p. 271), "the bulk of the nervous system must be patterned without the aid of functional adjustment," or, "Development in many instances . . . . is remarkably independent of function, even in . . . [the] sense [of] . . . function as a general condition necessary to healthy growth." Twenty years later, Sperry (1971, p. 32) continued to observe: "In general outline at least, one could now see how it could be entirely possible for behavioral nerve circuits of extreme intricacy and precision to be inherited and organized prefunctionally solely by the mechanisms of embryonic growth and differentiation." Sperry was not alone in expressing a genetically predeterministic conception of neural and behavioral epigenesis. Viktor Hamburger, perhaps the foremost student of Nobel laureate Hans Spemann, echoed Sperry's beliefs on several occasions, which, to his credit, he later ameliorated:

> The architecture of the nervous system, and the concomitant behavior patterns result from self-generating growth and maturation processes that are determined entirely by inherited, intrinsic factors, to the exclusion of functional adjustment, exercise, or anything else akin to learning. (Hamburger, 1957, p. 56, reiterated in toto in 1964, p. 21)

With noted authorities on the development of the nervous system making such statements in books and articles apt to be read by biologically oriented psychologists, it is not surprising that a genetically predeterministic view entered into psychology, especially a psychology trying to recover its balance from accusations (of the other error) of environmentalism. One of the values of a systems view of development is the explicit utilization of both genetic and experiential influences, not merely a nervous and often empty lip service averring that both are surely necessary.

THE TRIUMPH OF PROBABILISTIC EPIGENESIS OVER PREDETERMINED EPIGENESIS

As noted in an earlier chapter, in 1970 I described an extant dichotomy in conceptualizing individual development as the predetermined and probabilistic epigenesis of behavior. The former saw a solely genetically inspired structural maturation as bringing about function in a unidirectional fashion, whereas the latter envisaged bidirectional influences between structure and function. The range of application of the probabilistic conception did not seem very broad at the time. Later (Gottlieb, 1976a), I explicitly added the genetic level to the scheme so that the unidirectional predetermined conception was pictured as genetic activity → structure → function in a nonreciprocal pathway, whereas the probabilistic notion was fully bidirectional (genetic activity ↔ structure ↔ function). Now that spontaneous neural activity as well as behavioral and environmental stimulation are known to play roles in normal neural development, and that sensory and hormonal influences can trigger genetic activity, the correctness and broad applicability of the probabilistic notion are undeniable and widely confirmed. In this sense, the probabilistic conception of epigenesis has triumphed over the predetermined view.

Building on the probabilistic notion, more recently (Gottlieb, 1991b, 1992) I have presented a simplified scheme of a systems view of psychobiological development that incorporates the major points of von Bertalanffy's, Weiss's, and Wright's thinking on the subject, and adds some detail on the organism–environment level that is necessary for a thoroughgoing behavioral and psychobiological analysis. Whatever merit this way of thinking about development may have certainly must be traced to the pioneering efforts of psychobiological theoreticians such as Z.-Y. Kuo (summarized in 1976), T. C. Schneirla (1960), D. S. Lehrman (1970), and Ashley Montagu (1945, 1977). At present, the probabilistic, bidirectional conception is being used both implicitly and explicitly by a number of more recent psychobiologically oriented theorists (e.g., Cairns, Gariépy, & Hood, 1990; Edelman, 1988; Ford & Lerner, 1992; Griffiths & Gray, 1994; Hinde, 1990; Johnston & Hyatt, 1994; Magnusson & Törestad, 1993; Oyama, 1985).

As shown in Fig. 8.5, I have reduced the levels of analysis to three functional organismic levels (genetic, neural, behavioral) and an environmental level subdivided into three components (physical, social, cultural).[3] While those of us who work with nonhuman animal models stress the influence of the physical and social aspects of the environment, those who work with humans prominently include cultural aspects as well. The criticism that one hears most of this admittedly simple-minded scheme is

---

[3]Gariépy (1995) has correctly pointed out that psychological functioning as such is not included in the four levels of systems diagram (Figs. 6.1 and 8.5). The reason for that omission is that psychological functioning or mediation (perception, thinking, attitudes, love, hate, etc.) must be inferred from analysis at the overt level of behavior and the environment, as made clear by the notion of methodological behaviorism introduced by E. C. Tolman in 1932. In this sense, all psychologists are methodological (not theoretical) behaviorists (cf. Brunswik, 1952).

not that it is overly simple but, rather, that it is too complex, not only with too many influences but with too many influences running in too many directions. In short, a developmental systems approach is sometimes alleged to be unmanageable and just not useful for analytic purposes. Perhaps unfortunately, it does represent reality as we now understand it, so it depicts individual development at a suitable level of complexity that does justice to the actualities of developmental influences. In the same spirit, at the conclusion of their review of genotype and maternal environment, Roubertoux, Nosten-Bertrand, and Carlier (1990, p. 239) observed:

> The effects constitute a very complex network, which is probably discouraging for those who still hope to establish a simple relation between the different levels of biological organization, and particularly the molecular and the behavioral. The picture is indeed more complicated.

SUMMARY FEATURES OF A DEVELOPMENTAL-PSYCHOBIOLOGICAL SYSTEMS VIEW

In summary, in its finished form, the developmental-psychobiological systems approach involves a temporal description of activity at four levels of analysis (genetic, neural, behavioral, environmental) and the bidirectional effects of such activity among the four levels. When the related notions of bidirectionality and probabilistic epigenesis were first put forth, they were largely intuitive. They are now established facts. Given the experimental-embryological heritage of all systems views, two further assumptions or propositions are warranted. Because of the early equipotentiality of cells and the fact that only a small part of the genome is expressed in any individual (Gottlieb, 1992), what is actually realized during the course of individual psychological and behavioral development represents only a fraction of many other possibilities (also see Kuo, 1976, on this point). Finally, a developmental systems view entails the notion of equifinality, that is, the likelihood of variation in pathways to common developmental endpoints.

BROADER IMPLICATIONS OF A DEVELOPMENTAL-PSYCHOBIOLOGICAL SYSTEMS VIEW

Although there is considerable evidence for the vertical as well as the horizontal bidirectionality of influences among the four levels of analysis depicted in Fig. 8.5 (environment, behavior, neural activity, genetic expression), the completeness of the top-down flow (from environment to genetic expression) has not yet been widely understood and appreciated in devel opmental psychology. Waddington's (1957, p. 36, Fig. 5) unidirectional understanding of genetic canalization has been the predominant approach for many years and is still promoted in some quarters of developmental psychology (Fishbein, 1976; Kovach & Wilson, 1988; Lumsden & Wilson, 1980; Parker & Gibson, 1979; Scarr-Salapatek, 1976).

Because the influence of environmental factors on genetic expression is presently being pursued in a number of neuroscience and neurogenetic laboratories, there is now considerable evidence to document that genetic activity is responsive to the developing organism's external environment. In an early example, Ho (1984) induced a second set of wings on fruitflies by exposing them to ether during a certain period of embryonic development; the ether altered the cytoplasm of the cells and thus the protein produced by the DNA-RNA-cytoplasm coactional relationship. This particular influence has the potential for a nontraditional evolutionary pathway in that it continues to operate transgenerationally, as do the effects of many drugs and other substances (Campbell & Perkins, 1988). As noted in chapter 6 and earlier in this chapter, there are now so many empirical demonstrations of external sensory and internal neural events that both excite and inhibit gene expression that the phenomenon has been labeled *immediate early gene expression* (e.g., Anokhin et al., 1991; Calamandrei & Keverne, 1994; Mack & Mack, 1992; Rustak et al., 1990).

In contrast to the usually unidirectional bottom-up flow still prominent in developmental psychology, at the behavior-environment level of analysis, bidirectionality was prominently recognized as early as J. M. Baldwin's (1906) "circular reaction," Vygotsky's (van der Veer & Valsiner, 1991) emphasis on the person's coactions with the person's cultural worlds, and William Stern's (1938) personology or person-*Umwelt* relatedness, among many other more recent examples (Fischer, Bullock, Rotenberg, & Raya, 1993; Ford & Lerner, 1992).

It is interesting to note that in a recent analysis of the recognition of bidirectional influences in theoretical accounts of biology, psychology, and sociology, although psychological theory recognizes vertical bidirectionality at the environment-behavior level and micro to macro unidirectional flow at the gene to neural level, sociological theory predominantly sees unidirectional vertical influences at the environment-behavior level and a consequent lack of persons affecting their social and cultural worlds (Shanahan, Valsiner, & Gottlieb, 1997). Indeed, Shanahan et al. concluded that although examples of bidirectionality can be found across disciplines, unidirectional thinking is still quite common. It is only recently that biologists have found the macro to micro flow empirically justified, and this top-down influence has not yet taken hold in psychology as a whole (for an exception in developmental psychopathology, see Cicchetti & Tucker, 1994). Sociologists, on the other hand, have not yet widely embraced the micro to macro flow of influences at the behavior-environment level.

PROBABILISTIC EPIGENESIS IN DEVELOPMENTAL PSYCHOLOGY

The probable nature of epigenetic development is rooted in the reciprocal coactions that take place in complex systems, as indicated in Fig. 8.5 in this chapter and Fig. 6.1 in chapter 6.

Since the overthrow of biological preformation in favor of epigenesis in the 19th century, it has been recognized that development takes place sequentially and is therefore an emergent phenomenon. And since the advent of experimental embryology in the late 19th century, it is an accepted fact that cellular and organismic development occurs as a consequence of coactions at all levels from the genes to the developing organism itself. With the gradual realization that influences in developmental systems are fully bidirectional and that genes do not, in and of themselves, produce finished (i.e., mature) traits, the predetermined concept of epigenesis has receded from all but a few viewpoints in biology and psychology (cf. Scarr, 1993). Epigenesis is now defined as increased complexity of organization: the emergence of new structural and functional properties and competencies as a consequence of horizontal and vertical coactions among the system's parts, including organism-environment coactions. I hope developmental psychology will eventually embrace the notion of causality as coaction. As noted earlier, the multiple pathways to the same developmental endpoint are represented well in the concept of equifinality.

As concluded by Shanahan et al. (1997), probabilistic epigenesis is in accord with Baldwin's (1906) understanding of developmental phenomena. The stochastic nature of developmental phenomena ultimately derives from the fact that there is a range of possible responses at any given level. Thus, responses to tension can vary within levels; given that responses to stress occur in highly related sets of behavior (i.e., they are organized), there will be variability in the overall patterns between levels. London's (1949) argument for the "behavioral spectrum" exemplifies the concern for a range of responses. From this perspective, developmental phenomena cannot be represented so as to imply subsequent derivations, although they can suggest classes of outcomes. This notion is captured well by Fischer's theory of cognition, as he adopts the principles of adaptive resonance theory to explain the generation of multiple cognitive forms in ontogeny (Fischer et al., 1993).

Thus, the hallmarks of probabilistic epigenesis — bidirectionality and indeterminacy — are being ever more widely used in developmental psychology, even if they are not yet majority opinions among psychological theorists who are not steeped in our own history of conceptualizing behavior-environment relations or who have yet to grasp the recent empirical breakthroughs in our understanding of biological development.[4]

SUMMARY AND CONCLUSION

Developmental thinking began in the early 1800s coincident with the triumph of epigenesis over the concept of preformation. Although practiced only at the descriptive level in this early period, it led to the insight that to understand the origin of any phenotype it is necessary to study its development in the individual. Late in the 1800s, developmental description was superceded by an experimental approach in embryology, one explicitly addressed to a theoretical understanding and explanation of developmental outcomes. A field or systems view was born when the results of Hans Driesch's experiments made it necessary to conceptualize embryonic cells as harmonious-equipotential systems. Steering a careful path between mechanical-reductive and vitalistic-constructive viewpoints, in the 1930s Ludwig von Bertalanffy formalized an organismic systems view for experimental embryology, which was later worked out in more formal detail by the embryologist Paul Weiss and the physiologically oriented population geneticist Sewall Wright. At present, a systems view of psychobiological development has begun to take hold in developmental psychology, developmental neurobiology, and behavior genetics (Gottlieb, Wahlsten, & Lickliter, in press). Thus, although there are dissenters, a psychobiological systems view seems workable and useful in understanding human as well as nonhuman animal psychological development.

It is quite rewarding to those who work with nonhuman animals to note that Ford and Lerner (1992) explicitly advocate the utility of a systems concept for developmental psychologists who work with human beings. As noted earlier, similar points of view have been put forward by psychobiologically oriented developmentalists such as Cairns et al. (1990), Edelman (1988, 1992), Griffiths and Gray (1994), Hinde (1990), Johnston and Hyatt (1994), Magnusson and Törestad (1993), and Oyama (1985). This represents a realization of the pioneering theoretical efforts of Z.-Y. Kuo (1976), T. C. Schneirla (1956, 1960), D. S. Lehrman (1970), and A. Montagu (1945, 1977). Because a developmental systems view dates to as early as Hans Driesch's theorizing about his embryological experiments in the 1890s, one cannot call it a "paradigm shift," but certainly it is something relatively new in the field of developmental psychology.

---

[4]It must be a cause of puzzlement and consternation to students and others when leading "developmental" behavior geneticists honestly state that their theoretical approach does not concern the paths from gene to behavior (e.g., Scarr, 1997).

9
Toward a New Developmental and Evolutionary Synthesis

Just as I was wrapping up this work I became aware of an odd convergence — from cognitive psychology and from developmental biology — that stimulated me to think about what the future might hold. Strange as it may seem, prominent authors from the fields of cognitive psychology and developmental neurobiology were homing in independently on the same conclusion; namely, that the idea of a sheerly genetic determination of form and function is too narrow and should be replaced by the broader concept of an embryological "morphogenetic field," which includes genes but goes beyond them to include cell–cell coactions as well. A sheerly genetic determination would thus be replaced by a cellular neurobiological determination. This is an improvement, of course, but it ignores the feeddownward influences from the environment and behavior (the two top levels in Fig. 8.5). It is still a strictly bottom-up approach to neurobiological and psychological development.

I first describe and comment on the specific proposals from the cognitive psychologists and then do the same for the developmental biologists.

DEVELOPMENTAL COGNITIVE PSYCHOLOGY

The cognitive psychologists want to redefine *innate* in light of their appreciation of the relevance of biology, particularly neurobiology, to developmental psychology. To my mind, they have misappropriated the concept of innateness in the service of their discovery of neurobiology. They don't have a theory of development unless they can use the innateness concept, which, in their hands, represents a too-narrow appreciation of the contribution of neurobiology to behavior/psychology, in the sense of biologically inspired predispositions or biases (Spelke & Newport, in press) or biological constraints (Elman et al., 1996). The latter define *innate* "to refer to putative aspects of brain structure, cognition or behavior that are the product of interactions internal to the organism . . . this usage of the term does not correspond to genetic or coded in the genes." Their Table 1.2 specifies that innate outcomes are the consequence of molecular and cellular interactions.

Because Spelke and Newport (in press) use the usual narrow definition of experience (obvious contributory organism–environment encounters) and not the broader definition of functional activity advocated here, they actually end up with the old notion of innate (= not only independent of frank learning but independent of experience). For Spelke and Newport, the resolution of the age-old innate (nature)–acquired (nurture) dichotomy is to accept it as valid, not in the sense that innate = genetically determined but, rather, innate = biologically predisposed, a higher order, morphogenetic field concept. The resolution I have opted for in this volume is to accept that certain developmental outcomes are species-typical or species-specific, adaptive, and responsive to a narrow class of stimulation in the absence of prior exposure to these configurations (independent of frank learning but not independent of experience, broadly defined). Who would have dreamed that squirrel monkeys' innate fear of snakes could derive from their experience with live insects (Masataka, 1994)? Or that chicks' perception of mealworms as food items depends on their having seen their own toes move (Wallman, 1979)?

Therefore, in the place of frank associationistic (S-R, S-S) learning, I hold open the likelihood of nonobvious experiential background factors as indispensable contributors to development. In this scheme, so-called biological predispositions or constraints are seen in a broader coactional context.

To my mind, it is only one step up from genetic determination to cite cellular-biological predispositions as sufficient causes of psychological outcomes. There remain nonobvious organism–environment experiences and two-way traffic in the developmental-psychobiological system whether one cares to accept them or not. That is the appropriate resolution of the nature–nurture dichotomy, to my way of thinking. Given the empirical demonstrations of the nonobvious prenatal experiential canalization of species-specific behavior described earlier, and the sensory stimulation of gene expression reviewed earlier, biologically encapsulated definitions of innate or instinctive predispositions or constraints (Elman et al., 1996) are no longer tenable. This is all the more true for human cognitive development, given that all of the sensory systems are capable of function prior to birth in humans (Gottlieb, 1971b). Gene expression is being activated by sensory stimulation, as well as other factors, in the fetal period. Behavioral embryological approaches to human development may also reveal organism–environment coactions that are relevant to the psychological adaptations of the infant (e.g., prenatal auditory experience leading to the newborn's selective auditory response to its mother's voice: DeCasper & Spence, 1986). Because Elman et al. are interested in language development, it is a pity that they do not recognize the prenatal auditory experience of speech in the human fetus as a developmental-psychobiological constraint. This issue is renewed more broadly by Locke (1993) and by Turkewitz (1988).

DEVELOPMENTAL AND EVOLUTIONARY BIOLOGY

A genuine paradigm shift seems to be gradually taking hold in the study of developmental biology, and it has implications for the study of evolution. Although I recognize the shift as momentous, I feel that it does not go quite far enough, most likely because of limitations in training, self-learning, and expertise that we all share.

For many years the evolutionary changes in physiology, morphology, behavior, and psychology have been ascribed to changes in gene frequencies in populations of interbreeding individuals, meaning that a change in the genetic composition of a population was believed to be required for an evolutionary change in a phenotype. This view of genetic change → phenotypic change jumps right over development. A few brave souls, who were significantly out of step with the times, speculated that changes in genes brought out opportunities for novel coactions during embryonic development, and it was the consequence of these new developmental coactions that brought about changes in morphology that were recognized as evolutionary changes (e.g., de Beer, 1930, 1958; Garstang, 1922; Goldschmidt, 1933, 1952). That view has now been elaborated, with considerably more empirical support, in the hands of a small number of biologists who see the embryological "morphogenetic field" as the appropriate focus of both developmental and evolutionary understanding (e.g., Edelman, 1992; Gilbert et al., 1996; Holliday, 1990).

This is a substantial change in the focal level of analysis, moving from a primary focus on genes to cell–cell coactions, which of course includes the genes but goes several levels beyond (above) them. Although the coactions are usually seen as vertically and horizontally bidirectional, the biologists, as do the cognitive psychologists, stop short of including behavior and the external environment as part of their developmental systems analysis (Fig. 8.5). An important consideration in understanding this shortcoming is that no one is yet competent to move comfortably (i.e., knowledgeably) between the four levels in Fig. 8.5. This is why broader interdisciplinary and multidisciplinary collaborations will be necessary in the future if behavior and psychological functioning are to become a genuine part of the new developmental and evolutionary synthesis.

Given the present context (the newness of the emancipation from a gene-centered focus to the cellular level in developmental cognitive psychology, and in evolutionary and developmental biology), it is perhaps understandable that the present new synthesis must remain for the time being a strictly bottom-up approach as far as a neural, behavioral, and psychological understanding are concerned. When George Miklos (1993, p. 851) wrote, "Control of the genome means control of development, and control of development means control of behaviors," he declared a strategy for an experimental program of research that, in my opinion, will one day have to be joined by a program of study moving in the opposite direction if we are to achieve an adequate understanding of the development and evolution of genes, nervous systems, and behaviors. It is now known that light induces changes in the circadian rhythms of behavioral activity in fruit flies by affecting gene expression of the proteins that are implicated in setting the circadian clock (Lee, Parikh, Itsukaichi, Bae, & Edery, 1996; Myers, Wager-Smith, Rothenfluh-Hilfiker, & Young, 1996). That should alert even the most conservative bottom-up person to the reality of top-down influences during the normal course of development.

To try to lend a hand in what one day will surely be a magnificent synthesis, I offer the following brief recap of a macrodevelopmental, behaviorally driven scheme of evolutionary change, a top-down approach meant to mesh with the bottom-up approach (Gottlieb, 1992).

THE RELATIONSHIP OF DEVELOPMENT TO EVOLUTION

The insightful concept that changes in individual development are the basis for evolution was raised originally by St. George Mivart (1871) in his book *On the Genesis of Species*. William Bateson (1894) favored the idea that phenotypic variation was developmentally inspired, but little was done to further work out the details until Walter Garstang (1922) and Gavin de Beer (1930, 1958) delivered their respective coups de grâce to Ernst Haeckel's recapitulation doctrine, and, from another side entirely, Richard Goldschmidt (1933, 1952) hypothesized that changes in early embryonic development would be necessary for evolution to occur. Although Garstang and de Beer were interested in showing the importance of various kinds of ontogenetic changes to evolution generally, Goldschmidt, having become convinced of the impossibility of neo-Darwinian microevolution producing a new species, had come to view developmental macromutation as essential to the production of the large differences necessary for speciation.

The foregoing scientists were the principal ones to establish the developmental basis of evolutionary change. The view that I wish to propose in this chapter builds on their pioneering insight, but it is different in a very important way. To be specific, Garstang, de Beer, and Goldschmidt, in agreement with the proponents of the modern synthesis, believed that a genetic change or mutation is necessary to bring about the developmental changes that lead to evolution. The point I wish to advocate is that there is so much untapped potential in the existing developmental system (including the genes) that evolution can occur without changing the genetic constitution of a population. Such changes may eventually lead to a change in genes (or gene frequencies), but evolution will have already occurred at the phenotypic level before the genetic change occurs. According to the present viewpoint, genetic change is a secondary or tertiary consequence of enduring behavioral changes brought about initially by supragenetic alterations of normal or species-typical development.

For those readers that were taught (or learned on their own) that changes in gene frequencies drive evolutionary changes in morphology, it may come as a surprise that there is no relationship between morphological complexity and genome size. If evolution is entirely gene based, one would have thought it would take more genes to build more complex bodies, but that is not the case, as can be seen in Fig. 9.1. If there were a relationship between genome size and morphological complexity, you would see an increase in genome size moving from the lower left of the diagram to the upper right of the diagram in a more or less straight line.

For our present developmental psychobiological purposes, it is even more relevant to know that there is no relationship between the number of genes coding for protein and the number of neurons in the nervous system. As shown in Table 9.1, although mice and humans each have approximately 70,000 coding genes, mice have approximately 40 million neurons and humans have about 85 *billion* neurons. At the invertebrate level, while the round worm (*Caenorhabdhitis elegans*) and the fruit fly (*Drosophila melanogaster*) have around 12,000–14,000 coding genes, the roundworm has 302 neurons and the fruit fly 250,000 neurons. Thus, there is good reason to seek the answer to evolution above the level of the genes, in the total developmental system.

THE INDUCTION OF BEHAVIORAL NEOPHENOTYPES

In a book closing out his underappreciated but otherwise illustrious research career as a broadly based developmental scientist, Zing-Yang Kuo

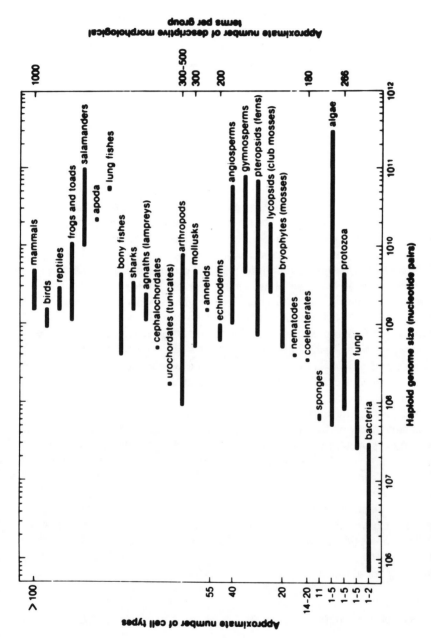

FIG. 9.1. The C-value paradox: absence of a relationship between genome size and morphological complexity. The bars show the ranges of genome sizes for various categories of organisms. The somewhat subjective ordering of categories is from morphologically most simple at the bottom to most complex at top. Two estimates of complexity are given: Approximate numbers of cell types in the body of some groups are indicated on the left vertical axis, and approximate numbers of morphological descriptive terms for certain groups are indicated on the right vertical axis. C-value data from Sparrow, Price, and Underbrink (1972). Figure from Raff and Kaufman (1983). Reprinted by permission.

TABLE 9.1
Approximate Number of Genes and Neurons in Different Lineages

	Genes	Neurons
Chordates		
Mus musculus	70,000	40 million
Homo sapiens	70,000	85 billion
Nematodes		
Caenorhabdhitis		
elegans	14,000	302
Arthropods		
Drosophila		
melanogaster	12,000	250,000

Note. From Miklos and Edelman (1996). Reprinted by permission. The actual number of neurons in *C. elegans* is known to be 302. Otherwise, the other figures in this table are approximations.

(1976) coined the term *behavioral neophenotype* to refer to momentous behavioral changes or deviations from normality that could be brought into existence by altering the usual conditions of an animal's early development or experience. Kuo's purpose for advocating the creation of behavioral novelties was to show that species-typical behavior was rather more highly modifiable than anyone believed and not rigidly or narrowly fixed by genetic constraints. To make his point, Kuo did such things as "create" a male dog that had no reproductive interests in female dogs in heat and, further, actively prevented other males from engaging in sexual behavior with such females. Kuo chose to tamper with reproductive behavior to make his point all the more telling: The usual or normal behavioral predilection for male dogs to copulate with females in heat (and thereby perpetuate the species) is a result of their having been exposed to usual or normal developmental conditions, not to instincts dictated solely by their genetic endowment. Kuo's argument that the establishment of behavioral neophenotypes by altering developmental circumstances should be one of the major aims of experimental animal psychology has not caught on because it seemed to many to be nonbiological. I hope that by supplying a broader rationale the significance of behavioral neophenotypes will be better appreciated.

If my presentation of the evidence to this point has been persuasive, the reader should be convinced that internal and external coactions during individual development create the resulting phenotype. This will, of course, be just as true for behavior as for anatomy or physiology. According to the viewpoint being developed in this chapter, the ease or difficulty of creating behavioral neophenotypes, and the directions of most ready behavioral change, would allow us to assess the immediately present evolutionary potential of a species. Naturally, we will expect to find that some species possess much greater immediate behavioral malleability or plasticity than other species (e.g., wood ducklings vs. mallards), and that in itself will be informative about the pace and range of immediate evolutionary potential in those species.

But I am getting a little bit ahead of my story. I am certainly not being original in suggesting that behavioral innovations lead the way to evolutionary change. A number of biologists of various persuasions have resuscitated Lamarck's notion of the centrality of behavioral change to evolution (e.g., Bonner, 1983; Hardy, 1965; Larson, Prager, & Wilson, 1984; Leonovicová & Novák, 1987; Mayr, 1982; Piaget, 1978; Plotkin, 1988; Reid, 1985; Sewertzoff, 1929; Wyles, Kunkel, & Wilson, 1983). What is new and not yet widely appreciated is the supragenetic means (neophenogenetic pathway described next) by which normal or usual development can be altered so as to produce a behavioral neophenotype that is likely to lead to evolutionary change. (In an excellent chapter on this topic, P. Bateson [1988] has come to much the same conclusion.)

In essence, as is widely recognized, what needs to happen to bring about evolution is the production of animals that live differently from their forebears. Living differently, especially living in a different place, will subject the animals to new stresses, strains, and adaptations that will eventually alter their anatomy and physiology (without necessarily altering the genetic constitution of the changing population). The new situation will call forth previously untapped resources for anatomical and physiological change that are part of each species' already existing developmental adaptability. At some time further down the road it is possible the genetic makeup of the evolving population may change, but by the time that happens (if it does) the new behavioral, anatomical, and physiological changes will already be in place. The neophenogenetic pathway for evolutionary change is thus seen as (a) an alteration of development leading to a significant change in behavior, followed by (b) a change in morphology, and, eventually, possibly (c) a change in genetic composition of the population. Consistent with their view of the strictly genetic determination of the phenotype, adherents of the modern synthesis would consider that evolution occurred only if and when step c was achieved. From the present point of view, enduring transgenerational changes in behavior and morphology (i.e., phenotypic evolution) have occurred by step b, without the necessity of adding to, subtracting from, or otherwise changing the original genetic composition of the population. The present view holds that genes are part of a very flexible and highly adaptable developmental system, but that genes do not determine the features of the mature organism. Consequently, from this point of view, evolution involves changes in the developmental system (of which the genes are an essential part), but not necessarily changes in the genes themselves. For instance, it is entirely consistent with the present proposal that alterations in development may cause genes to become active in the developmental process that were heretofore quiescent. It is well accepted among developmental geneticists that only a very small portion of the genome is expressed during individual development, so there is always present a large untapped genetic resource that can be brought to surface under abnormal (non-species-typical) developmental circumstances, whether internal or external to the organism. The behavioral neophenogenetic pathway of evolutionary change is depicted in Table 9.2.

DETERMINANTS OF BEHAVIORAL PLASTICITY

The present theory lays great store in the malleability or adaptability of organisms, especially the higher vertebrates (birds and mammals — more on this latter point later). The creation of behavioral neophenotypes is necessarily dependent on the existence of some degree of behavioral

TABLE 9.2
Three Possible Stages in Evolutionary Pathway Initiated by Behavioral Neophenotype

I: Change in Behavior	II: Change in Morphology	III: Change in Genes
First stage in evolutionary pathway: Change in ontogenetic development results in novel behavioral shift (behavioral neopheotype), which encourages new environmental relationships.	Second stage of evolutiory change: New environmental relationships bring out latent possibilities for morphological–physiological change. Somatic mutation or change in genetic regulation may also occur, but a change in structural genes need not occur at this stage.	Third stage of evolutionary change: Resulting from long-term geographic or behavioral isolation (separate breeding populations). It is important to observe that evolution has already occurred phenotypically before stage III is reached. Modern neo-Darwinism, however, does not consider evolution to have occurred unless there is a change in genes or gene frequencies.

Note. From Gottlieb (1992). Reprinted by permission.

plasticity or adaptability. Thus, the determinants of behavioral plasticity are an important consideration. One key limiting component of plasticity is the nervous system, particularly the brain, and the other is the developing organism's early experiences. These two components are in lockstep: Larger brained species can make more of their early experiences, and early experiences affect the maturation and size of the brain. Thus, the most conspicuous developmental route to increasing behavioral plasticity and creating behavioral neophenotypes is through early experiential alterations (including nutrition) that have positive effects on enhancing the maturation of the brain.

Beginning in the 1950s, developmental psychobiologists began in earnest to study the influence of early rearing experiences on enhancing the nervous system and later exploratory behavior and problem-solving ability, the latter two interrelated forms of behavior and psychological functioning being of most relevance in engendering the sort of evolutionary progression described previously and depicted in Table 9.2. For the present purposes, we are most interested in the developmental conditions that produce the sorts of behavioral plasticity that would enhance the likelihood of an individual (a) being able to survive by behavioral means in a drastically changed environment or (b) whose behavior would be likely to bring it into a new environment, thus precipitating the anatomical and physiological changes in stage II in Table 9.2.

Led by the pioneering experiments of Seymour Levine (1956) and Victor Denenberg (1969), a large number of studies showed that the unusual, perhaps stressful, experience of subjecting young rodents to handling by human beings during early development resulted in producing relatively stress-resistant animals, ones that would be capable of exploration (instead of freezing) and adaptive learning when faced with a completely strange and unfamiliar environment in adulthood. The research of Levine (1962) and his collaborators showed that the axis between the adrenal and pituitary glands was enhanced by the handling experience, and this anatomical-physiological change was correlated with the handled animal being able to tolerate greater stress in adulthood. As shown by Denenberg (1964) and his colleagues, the handling experience had to occur early in development if it was to be effective. Animals subjected to the same experience at older ages did not benefit from the experience, as indicated by later tests of resistance to stress and of exploratory behavior. A particularly important feature of these experiments is that the effect wrought in one generation persists into the next generation, even though the genes have not been altered (Denenberg & Rosenberg, 1967).

In a series of experiments with a rather different purpose, Hymovitch (1952) showed in a definitive manner how variations in early experience are crucial to later *problem solving* in adulthood. He reared young rats under four conditions and then later tested them in the Hebb–Williams maze. The animals were housed individually in:

1. A stovepipe cage (which permitted little motor or visual experience).
2. An enclosed running or activity wheel (which permitted a lot of motor activity but little variation in visual experience).
3. A mesh cage that restricted motor activity but allowed considerable variation in visual experience as it was moved daily to different locations in the laboratory.
4. The fourth group of animals contained 20 animals that were reared socially in a so-called free environment box that was very large (6 × 4 ft) compared to the other conditions, and was fitted with a number of blind alleys, inclined runways, small enclosed areas, apertures, and so on, that offered the rats a wide variety of opportunities for motor and visual exploration and learning in a complex physical environment.

The animals lived in these four environments from about 27 days of age to 100 days of age, at which time testing in the Hebb–Williams maze was completed.

Although rearing in the stovepipe and the enclosed running wheel led to the same level of poor performance, rearing in the mesh cage and the free environment led to the same level of good performance over 21 days of testing in the Hebb–Williams maze. All of the groups also showed the same level of improvement over the 3 weeks of testing, so the animals reared in the mesh cages and free environment began functioning at a superior level early in testing.

Next, in order to determine whether it was the early experience in each environment that made for the differences between the groups, Hymovitch repeated the experiment with four groups of animals that differed in *when* they had the free-environment or stovepipe experience: One group had the free-environment experience from 30 to 75 days of age and then were placed in the stovepipe for 45 days; a second group had the stovepipe experience from 30 to 75 days and then had the free-environment experience for 45 days; a third group remained in the free environment throughout the experiment; and a fourth group remained in their normal laboratory cages throughout the experiment (these would be the most thoroughly or consistently deprived from the standpoint of motor and visual experience).

The animals that experienced the free environment early and the stovepipe later in life performed just as well as the animals that remained in the free environment throughout the experiment. The crucial finding is that the animals who experienced the stovepipe environment early and the free environment later in life performed as poorly as the animals that remained in their normal cages throughout the experiment (the most deprived group). It is important to note that these differences in problem-solving ability were not in evidence when Hymovitch challenged the rats with a simpler, alley maze, more like the ones that were in wide use in most animal learning laboratories at the time. It was only when they were challenged by the much more difficult Hebb–Williams series of problems that the differences in problem-solving ability were in evidence.

Forgays and Forgays (1952) undertook to replicate Hymovitch's important findings and also to determine (a) whether the "playthings" in the free environment were crucial and (b) why the mesh-cage-reared animals did so well without direct experience of interacting with the multifarious objects in the free environment. They found indeed that the "playthings" (inclined planes, blind alleys, etc.) were essential to the superior performance of the free-environment animals and that the mesh-cage-reared animals only do as well when their cages are moved about frequently so that they visually encounter a considerable degree of varied environmental input, including the opportunity to watch the animals in the free environment with the playthings.

It was not long before these early experience studies were extended to other animals, including nonhuman primates, where social isolation and otherwise highly restricted, deprived rearing conditions were employed. Indeed, even in primates with relatively large brains, the normal or usual variety of experiences early in life was critical for the appearance of normal exploratory and learning abilities later in life. Deprived infants showed severe deficiencies in their later behavior (Harlow, Dodsworth, & Harlow, 1965). Just having a large brain is insufficient for the development and manifestation of the superior problem-solving skills characteristic of primates (Mason, 1968; Sackett, 1968). Thus, behavioral plasticity that is essential to behavioral neophenogenesis is dependent on variations in early experience as well as possessing a large brain.

The conditions that favor the appearance of a behavioral neophenotype are severe or species-atypical alterations in environmental contingencies early in life. These changed contingencies can arise in two ways in animals living in nature:

1. Some sort of physical or geographical change happens *to* (is forced on) the animal (a disruption of habitat, climatic change, and so on).
2. Probably more frequently, the migration of the animal into a somewhat different habitat based on normal exploratory behavior.

The large-brained animals that are more likely to withstand (1) and commit (2) are ones that have had not only traditional but nontraditional variations in their early experience. To put it the other way around, exposure to conservative or narrow social and physical environmental contingencies early in development will make animals less likely to withstand (1) and unlikely to perpetrate (2).

These predictions on evolutionary readiness, as it were, follow from the results of the early experience studies reviewed previously. There is a developmental dynamic that causes animals to prefer the familiar and thus to strive to reinstate earlier life situations or repeat versions of their early life experiences in adulthood. Consequently, animals that have had considerable variation in their early social and physical experiences will tend to seek out such variation in adulthood—just what is needed to heighten exploratory behavior and encourage novelty seeking! Although actual developmental experiments have not yet been done to show that animals (including humans) that have had considerable variation in their early experience will tend to seek out novel experiences as adults, there are two studies of adult mammals and birds that show that novelty is a psychological dimension of experience that can be abstracted such that animals so trained will consistently prefer to interact with novel rather than familiar objects or situations when given a choice (Honey, 1990; Macphail & Reilly, 1989). From the present theoretical standpoint, it would be most valuable to validate the developmental induction of novelty-seeking behavior in later life through the experience of considerable variation early in life.

Another way, albeit indirect, to test the behavioral neophenogenetic hypothesis about evolutionary readiness is to examine the exploratory behavior and rate of evolutionary change in large-brained versus smaller brained species. The cerebral component of behavioral neophenogenesis predicts a higher degree of exploratory behavior in large-brained species versus small-brained species and a consequent faster rate of evolutionary change in larger versus smaller brained species. The modes of behavioral neophenogenesis reviewed thus far are summarized in Table 9.3.

EXPLORATORY BEHAVIOR AND RATE OF EVOLUTIONARY CHANGE IN LARGE- VERSUS SMALL-BRAINED SPECIES

After J. Huxley (1957), Bernhard Rensch (1959), and other evolutionary biologists agreed on the pinnacle status of birds and mammals based on considerations of ontogenetic and behavioral plasticity, a comparative psychologist named Harry Jerison (1973) produced a monumental tome, *Evolution of the Brain and Intelligence*, in which he was able to show that

TABLE 9.3
Modes of Behavioral Neophenogenesis

Unusual (e.g., "handling") and enriched early experiences lead to:
1. Increased resistance to stress
2. Increased brain size
3. Enhanced exploratory behavior
4. Enhanced problem solving (learning ability) in adulthood
These would aid adaptation should (a) the organism's usual environment change drastically and (b) would also support the seeking out of new habitats in the absence of environmental change; (1) and (2) are often invoked as the initial stages of evolution.

Note. From Gottlieb (1992). Reprinted by permission.

birds and mammals are in a class by themselves as far as the evolution of brain:body ratio is concerned. As can be seen in Fig. 9.2, at any given body weight, birds and mammals have a higher brain weight than all species of lower vertebrates at the same body weight. Consequently, according to the ideas already developed in this chapter, we should expect to see, by and large, greater behavioral plasticity in birds and mammals than in lower

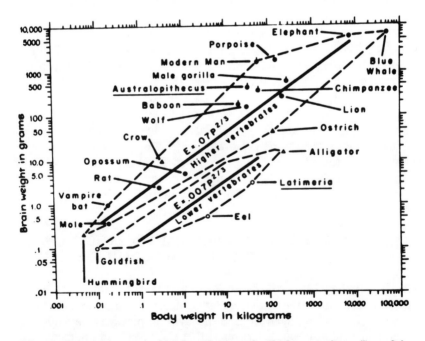

FIG. 9.2. Brain:body ratios of birds and mammals ("higher vertebrates") vs. fish, amphibians, reptiles ("lower vertebrates"). From Jerison (1969). Reprinted by permission from the University of Chicago Press.

vertebrates. As it happens, that prediction does accord rather well with learning ability, conceived of as a species' ability to show forms of learning above the level of conditioning: Sensory preconditioning and learned stimulus configuring are possible only in birds and mammals (Razran, 1971, p. 221).

From the present standpoint, one of the most interesting findings in an ambitious experimental study of exploratory behavior in a very large variety of vertebrates is the clear superiority of all mammalian forms tested (Fig. 9.3). In their study, Glickman and Sroges (1966) studied over 300 animals in over 100 different species, certainly the most extensive survey of exploratory behavior ever undertaken, so these results are most impressive from the standpoint of the consistency of mammalian superiority over the other vertebrate species.

It has been our contention that exploratory behavior – when a species is sufficiently plastic to initiate it – places the individual in a different niche facing different selective (adaptive) demands and thereby brings out latent morphological changes that then allow a genetically based evolutionary change to follow in its wake, as summarized in Table 9.2.

This sort of scheme allows evolution to proceed at a much more rapid rate than is the case when a species must await a severe environmental change or catastrophe such as oceans drying up or months of darkness, to prune individuals that are capable of adapting to the change from those that are not. Such extreme environmental changes are rare and widely spaced in time, compared to the frequency and tempo of evolution that could be instigated by behavioral neophenotypes. Consequently, given the relationship between large brains and behavioral plasticity, behavioral neopheno-

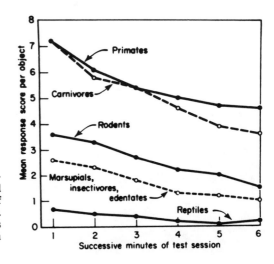

FIG. 9.3. The number of exploratory approaches made to novel objects by a large number of mammalian and reptilian species. From Glickman and Sroges (1966). Reprinted by permission from E. J. Brill.

genesis predicts that species with large brains should show evidence of a faster evolutionary pace than species with smaller brains. That prediction accords rather well with the finding of Wyles et al. (1983) of an almost perfect correlation between relative brain size and rate of anatomical evolution for a large number of vertebrate species.

As shown in Table 9.4, humans, the group with the largest relative brain size (Figure 9.2), show the fastest rate of anatomical evolution, with the larger and older hominoid groups ranking second in brain size and rate of evolution. What is perhaps most interesting is that the relatively recently evolved songbirds rank just below hominoids and well above other mammals and other birds on both relative brain size and rate of anatomical evolution. Finally, the classifications of "Other mammals" and "Other birds" show a relatively larger brain size than the lower vertebrates (lizards, frogs, salamanders) and a corresponding faster tempo of evolutionary change. Consonant with the present theory, but without the developmental component, Wyles and colleagues invoked behavioral innovation, as well as a large gene pool, as the major driving force for the observed differences in rates of anatomical evolution:

> Behavioral innovation refers to the nongenetic (or genetic) origin of a new skill in a particular individual, leading it to exploit the environment in a new way . . . . [The] nongenetic propagation of new skills and mobility in large populations will accelerate anatomical evolution by increasing the rate at which anatomical mutants of potentially high fitness are exposed to selection in new contexts. (Wyles et al., 1983, p. 4396)

STAGE II: CHANGE IN MORPHOLOGY WITHOUT CHANGE IN GENES

The present viewpoint takes advantage of the well-accepted fact that only a very small proportion of an individual's genotype participates in the

TABLE 9.4
Brain Size in Relation to Rate of Anatomical Evolution

Taxonomic Group	Relative Brain Size	Anatomical Rate
Homo	114	> 10
Hominoids	26	2.5
Songbirds	23	1.6
Other mammals	12	0.7
Other birds	4.3	0.7
Lizards	1.2	0.25
Frogs	0.9	0.23
Salamanders	0.8	0.26

Note. From Wyles, Kunkel, and Wilson (1983). Reprinted by permission.

developmental process. Thus, behavioral and morphological phenotypic changes can be immediately instigated by a change in an individual's developmental conditions. In our view, a change in developmental conditions activates heretofore quiescent genes, thus changing the usual developmental process and resulting in an altered behavioral or morphological phenotype. Consequently, stage II in the evolutionary pathway (Table 9.2) holds that the new environmental relationships bring out latent possibilities for morphological–physiological change in advance of the usual criterion of evolution: a change in structural genes or gene frequencies in the population (stage III). It is very exciting that this course of events corresponds to what is known about the correlation of morphological and genetic change in the evolutionary record. According to the best measurement techniques currently available, it would appear that morphological change antedates structural genetic and chromosomal change among the major vertebrate groups (Larson et al., 1984; Sarich, 1980, and references therein). Such a state of affairs is precisely compatible with the notion that behavioral neophenogenesis initiates a process of anatomical change that culminates only later in genic and chromosomal change. Because bringing extragenetic or supragenetic considerations into the evolutionary process, even in its introductory phases, is unorthodox, I would like to point out that this view can be integrated with population-genetic thinking and the modern synthesis (see next section), if the reality or plausibility of the early supragenetic stages I and II can be granted. Because those working on the problem of the apparent lack of synchrony in morphological and chromosomal evolution have themselves looked to behavior (specifically, social behavior) as the means whereby morphological change may be accelerated over chromosomal change in the course of evolution (Larson et al., 1984), the present author's contribution may be seen in the addition of the developmental dimension and its influence on the larger category of behavioral plasticity (exploratory behavior, problem solving) that is so essential to the early supragenetic phases of evolution. It is now acknowledged in many different quarters, both within and without the modern synthesis, that the time has come to include the role of individual development in evolution (e.g., P. Bateson, 1988; Futuyma, 1988; Gilbert et al., 1996; Goodwin, 1984; Ho & Saunders, 1982; Johnston & Gottlieb, 1990; Løvtrup, 1987; Oyama, 1985; Rosen & Buth, 1980; Thomas, 1971).

INTEGRATION OF INDIVIDUAL DEVELOPMENT INTO THE POPULATION-GENETIC MODEL

Finally, to more explicitly integrate the present theory into the modern synthesis, the evolutionary pathway described here (Table 9.2) is consistent with the idea championed by Parsons (1981), among others, that a behaviorally mediated ecological independence precedes reproductive isolation when populations within a species first begin to split off (speciate). That is, these authors see behaviorally mediated changes in habitat selection (stage II in my scheme), especially microhabitat preferences, as the first step in the pathway to eventual speciation (reproductive isolation). According to this view, when reproductive isolation eventually occurs, it is based on a later developing genetic incompatibility (stage III here) between originally homogeneous gene pools. According to the present scenario, the cause of the original behavioral divergence (e.g., preferences for different temperatures, humidities, light intensities, diets, oviposition sites, and mating sites in fruit flies) would be found in differences in the developmental histories (stage I) of the individuals showing the divergent behaviors. The stage I or developmental contribution to behavioral differences mediating speciation events has not yet been widely appreciated in the literature of evolutionary biology, even where novel shifts in behavior are seen as the key to speciation. That is to say, the causes of the novel shifts in behavior have received little attention as yet (Parsons, 1981, p. 230). The present scenario, specifically the induction of a behavioral neophenotype through a change in developmental conditions, offers an explanation for the novel behavior.

In summary, although the architects of the modern synthesis all agree that the mechanisms of evolution are mutation or genetic recombination, selection, migration, and eventual reproductive isolation, the present work describes how migration (invasion of new habitats or niches) may occur without mutation (or recombination) and selection first initiating a change in genes or gene frequencies. Both laboratory and field research indicate that reproductive isolation (i.e., incipient speciation) can occur without major genetic alterations (reviews in Bush, 1973; Singh, 1989, pp. 445–446). The usual scenario—theoretically speaking—is that a major genetic alteration is the necessary *first* step in speciation (Mayr, 1954). Bush's review indicates that behavior can lead the way to speciation with genetic changes coming into play only later on, which is consonant with the evolutionary model presented here. In this respect, it is noteworthy that Mayr (1988, Essay 28, pp. 541–542) now advocates a pluralistic view of the factors that instigate evolutionary change, one that might not entirely reject the present approach.

BOLDER SPECULATION

A bolder and much more radical proposal would hold out for the possibility of morphological change initiated by behavioral change alone (i.e., without a shift in the physical environment). Tradition is recognized as an important component of animal as well as human behavior (Bonner, 1980), so what would be required under this more radical evolutionary scenario would be a change in some behavioral tradition, especially one affecting the rearing experience of young animals in the process of growing up, so that the subsequent behavior of the developing animals would be altered without necessarily causing them to leave their usual physical environment or niche. So far, all of the behaviorally mediated evolutionary scenarios have assumed an ecological change. The more radical notion of a strictly behaviorally mediated morphological change leading to speciation without an ecological or environmental shift has not been put forward before, at least to my knowledge.

Lamarck's (1809/1984) behaviorally mediated morphological evolution was said by him to be stimulated by a physical environmental change or stressor. The controversial Lamarckian experiments by Steele (1979) involved a morphological response to an environmental change. The possibility of strictly behaviorally inspired morphological change in the absence of stage II (Table 9.2) has not been previously entertained. Based on the evidence and other considerations discussed in this monograph, I think it remains a theoretical possibility. For example, a change in behavior almost inevitably brings about a change in social interactions, and some authors theorize that changes in social interaction prompted the evolution of human intelligence (reviewed in Byrne & Whiten, 1988). But these scenarios are couched within the terms of the modern synthesis (genetic change brought about by natural selection leads to evolutionary change in intelligence). In the framework developed here, an evolutionary change in brain structure and function might well occur without a genetic change in the population. There is such a tremendous amount of currently unexpressed developmental potential that behavioral and anatomical evolution would seem possible without the necessity of new genetic variations produced by mutation and genetic recombination.

HOW CAN CHANGES ARISING IN ONE GENERATION PERSIST ACROSS GENERATIONS?

It is appropriate to ask how the new phenotypic changes can be preserved from one generation to the next if there has not been a mutation or new genetic recombination. The answer is that the transgenerational stability of new behavioral and morphological phenotypes is preserved by the repetition of the developmental conditions that gave rise to them in the first place. Because genes are a part of the developmental system and cannot make traits by themselves, this same requirement (repetition of developmental conditions) holds for the transgenerational perpetation of new phenotypes stemming from mutation and genetic recombination, even though that requirement often goes unrecognized or unspecified in the modern synthesis account of evolutionary change.

Although I am unable to propose a specific molecular mechanism whereby the organism's new experiences activate previously unactivated DNA (to get the expression of previously inactive genes), the present proposal obviously assumes such a mechanism. For example, something akin to the "homeobox" would fit the bill (Edelman & Jones, 1993; Gehring, 1987; Ingham, 1988). Some such mechanism is required for the Bolder Speculation to work. The activation of previously inactive genes must be occurring when, for example, avian oral epithelial cells grow a "mammalian" tooth under altered developmental conditions (Kollar & Fisher, 1980).

It seems clear to me that the Bolder Speculation applies to our own species: Earlier, we had suggested that *Homo erectus* could have possibly evolved into *Homo sapiens* through a dramatic change in rearing practices (Gottlieb, Johnston, & Scoville, 1982). It remains to be seen whether the behavioral activation of new anatomical features from a previously unexpressed developmental potential is more widely applicable, or whether the more usual scenario is a behaviorally initiated ecological shift bringing out latent morphological change (stage II in Table 9.2). In either event, a developmentally wrought behavioral change would play an important role in evolution.

CONCLUSION

The present theory of behavioral neophenogenesis is manifestly a theory of vertebrate evolution, particularly of the higher vertebrates (birds and mammals), where the role of early experience in enhancing brain size, learning ability, exploratory behavior, and resistance to stress has been experimentally demonstrated in a number of species. The impact of early experience on later behavior is not without import in invertebrates (e.g., Jaisson, 1975; McDonald & Topoff, 1985) and lower vertebrates, but such developmental studies are few in number, so the question remains more open on the role of behavioral neophenogenesis in the evolution of invertebrate and lower vertebrate forms (fish, amphibians, reptiles). However, our general concept of neophenogenesis (Johnston & Gottlieb, 1990), which is a more global developmental theory of phenotypic evolution, would seem to have broad application in the evolutionary arena.

Finally, it is gratifying to see the concept of probabilistic epigenesis being used so contructively in some quarters of infant development (Bertenthal, Campos, & Kermoian, 1994), cognitive psychology (Bidell & Fischer, 1997), song learning in birds (Logan, 1992; West, King, & Freeberg, 1994), and medical genetics (review by Strohman, 1993), as well as in penetrating critiques of theories of development in biology, psychology, and sociology (Ford & Lerner, 1992; Gariépy, 1995; Lerner & Kauffman, 1985; Logan, 1985; Michel & Moore, 1995; Shanahan et al., 1997; van der Weele, 1995). With a view toward what the future might hold with respect to theory and experiment in development and evolution, as I noted in the introduction to this chapter, there has been a recent movement in some quarters of cognitive development and in developmental and evolutionary biology to shift the focus of analysis from the genetic level to cell–cell interactions (the morphogenetic field). However, this is still a strictly bottom-up, internalist view of development (van der Weele, 1995) that ignores the contribution of embryonic/fetal behavior and the prenatal environment in constructing the organism. Thus, in order to truly move ahead in our theories and experiments, it behooves us to include the behavioral and environmental levels in a constructive, bidirectional view of epigenesis and its bearing on evolution, and to pursue such behaviorally and environmentally inspired inquiries into the prenatal stages of gene expression and neural development. Multidisciplinary training and interdisciplinary research will be necessary if the discipline of developmental psychobiology is to meet its promise of unifying biology and behavioral science (Michel & Moore, 1995).

To fill the enormous gap between molecular neurobiology and psychological function, it will also be necessary to have mutual respect for top-down as well as bottom-up approaches — it is not enough to understand how the brain develops from a sheerly internalist perspective.

References

Alberts, J. R. (1978). Huddling by rat pups: Multisensory control of contact behavior. *Journal of Comparative and Physiological Psychology*, *92*, 220–230.

Anokhin, K. V., Milevsnic, R., Shamakina, I. Y., & Rose, S. (1991). Effects of early experience on c-fos gene expression in the chick forebrain. *Brain Research*, *544*, 101–107.

Armstrong, R. C., & Montminy, M. R. (1993). Transsynaptic control of gene expression. *Annual Review of Neuroscience*, *16*, 17–29.

von Baer, K. E. (1966). *Über Entwickelungsgeschichte der Thiere: Beobachtung und Reflexion* (Part one). New York: Johnson Reprint Corporation. (Original work published 1828 by Bornträger, Königsberg)

Baldwin, J. M. (1906). *Thought and things: A study of the development and meaning of thought, or genetic logic*, vol. 1. *Functional logic or genetic theory of knowledge*. London: Swan Sonnenschein.

Banker, H., & Lickliter, R. (1993). Effects of early and delayed visual experience on intersensory development in bobwhite quail chicks. *Developmental Psychobiology*, *26*, 155–170.

Barbe, M. F. (1996). Tempting fate and commitment in the developing forebrain. *Neuron*, *16*, 1–4.

Bateson, P. (1988). The active role of behaviour in evolution. In M.-W. Ho & S. W. Fox (Eds.), *Evolutionary processes and metaphors* (pp. 191–208). London: Wiley.

Bateson, P. P. G. (1976). Specificity and the origins of behavior. *Advances in the Study of Behavior*, *6*, 1–20.

Bateson, W. (1894). *Materials for the study of variation, treated with especial regard to discontinuity in the origin of species*. New York: Macmillan.

de Beer, G. (1930). *Embryology and evolution*. Oxford: Clarendon Press.

de Beer, G. (1958). *Embryos and ancestors* (3rd ed.). Oxford: Clarendon Press.

Bellugi, U., Wang, P. P., & Jernigan, T. L. (1994). Williams Syndrome: An unusual neuropsychological profile. In S. H. Broman & J. Grafman (Eds.), *Atypical cognitive deficits in developmental disorders: Implications for brain functions* (pp. 23–56). Hillsdale, NJ: Lawrence Erlbaum Associates.

von Bertalanffy, L. (1950). *A systems view of man.* Boulder, CO: Western Press.

von Bertalanffy, L. (1962). *Modern theories of development: An introduction to theoretical biology.* New York: Harper. (Original work published 1933 in German)

Bertenthal, B. I., Campos, J. J., & Barrett, K. C. (1984). Self-produced locomotion. In R. N. Emde & R. J. Harmon (Eds.), *Continuities and discontinuities in development* (pp. 175–210). New York: Plenum.

Bertenthal, B. I., Campos, J. J., & Kermoian, R. (1994). An epigenetic perspective on the development of self-produced locomotion and its consequences. *Current Directions in Psychological Science, 3,* 140–145.

Bidell, T. R., & Fischer, K. W. (1997). Between nature and nurture: The role of human agency in the epigenesis of intelligence. In R. Sternberg & E. Grigorenko (Eds.), *Intelligence: Heredity and environment* (pp. 193–242). New York: Cambridge University Press.

Black, I. B. (1993). Environmental regulation of brain trophic interactions. *International Journal of Developmental Neuroscience, 11,* 403–410.

Bloch, V. (1976). Brain activation and memory consolidation. In M. R. Rosenzweig & E. L. Bennett (Eds.), *Neural mechanisms of learning and memory* (pp. 583–590). Cambridge, MA: MIT Press.

Bonner, J. T. (1980). *The evolution of culture in animals.* Princeton, NJ: Princeton University Press.

Bonner, J. T. (1983). How behavior came to affect the evolution of body shape. *Scientia, 118,* 175–183.

Born, D. E., & Rubel, E. W.(1988). Afferent influences on brain stem auditory nuclei of the chicken: Presynaptic action potentials regulate protein synthesis in nucleus magnocellularis neurons. *Journal of Neuroscience, 8,* 901–919.

Brunswik, E. (1952). *The conceptual framework of psychology.* Chicago: University of Chicago Press.

Bush, G. L. (1973). The mechanism of sympataric host race formation in the true fruit flies *(Tephritidae).* In M. J. D. White (Ed.), *Genetic mechanisms of speciation in insects* (pp. 3–23). Dordrecht, Holland: D. Reidel.

Byrne, R. W., & Whiten, A. (Eds.). (1988). *Machiavellian intelligence: Social expertise and the evolution of intellect in monkeys, apes, and humans.* Oxford: Oxford University Press.

Cairns, R. B., Gariépy, J.-L., & Hood, K. E. (1990). Development, microevolution, and social behavior. *Psychological Review, 97,* 49–65.

Cairns, R. B., MacCombie, D. J., & Hood, K. E. (1983). A developmental-genetic analysis of aggressive behavior in mice: I. Behavioral outcomes. *Journal of Comparative Psychology, 97,* 69–89.

Calamandrei, G., & Keverne, E. B. (1994). Differential expression of Fos protein in the brain of female mice is dependent on pup sensory cues and maternal experience. *Behavioral Neuroscience, 108,* 113–120.

Campbell, J. H., & Perkins, P. (1988). Transgenerational effects of drug and hormone treatments in mammals. *Progress in Brain Research, 75,* 535–553.

Carlier, M., Roubertoux, P., Kottler, M. L., & Degrelle, H. (1989). Y chromosome and aggression in strains of laboratory mice. *Behavior Genetics, 20,* 137–156.

Cavalier-Smith, T. (1985). Cell volume and the evolution of eukaryote genome size. In T. Cavalier-Smith (Ed.), *The evolution of genome size* (pp. 105–184). Chicester, England: Wiley.

Changeux, J.-P., & Danchin, A. (1976). Selective stabilization of developing synapses as a mechanism for the specification of neuronal networks. *Nature, 264,* 705–712.

Changeux, J.-P., & Konishi, M. (Eds.). (1987). *The neural and molecular bases of learning.* Chicester, England: Wiley.

Cheng, M.-F. (1979). Progress and prospects in ring dove: A personal view. *Advances in the Study of Behavior, 9,* 97–129.

Cicchetti, D. V., & Tucker, D. (1994). Development and self-regulatory structures of the mind. *Development and Psychopathology, 6,* 533–549.

Cierpal, M. A., & McCarty, R. (1987). Hypertension in SHR rats: Contribution of maternal environment. *American Journal of Physiology, 253,* 980–984.

Clark, N. M., & Galef, B. G. (1988). Effects of uterine position on rate of sexual development in female mongolian gerbils. *Physiology & Behavior, 42,* 15–18.

Corner, M. A. (1994). Reciprocity of structure-function relations in developing neural networks: The Odyssey of a self-organizing brain through research fads, fallacies and prospects. *Progress in Brain Research, 102,* 3–31.

Cramer, C. P., Pfister, J. P., & Haig, K. A. (1988). Experience during suckling alters later spatial learning. *Developmental Psychobiology, 21,* 1–24.

Crow, T. J. (1995). Constraints on concepts of pathogenesis: Language and the speciation process as the key to the etiology of schizophrenia. *Archives of General Psychiatry, 52,* 1011–1014.

Curran, T., & Morgan, J. I. (1987). Memories of *fos. BioEssays, 7,* 255–258.

Davidson, E. H. (1986). *Gene activity in early development.* Orlando, FL: Academic Press.

Davis, J. O., & Phelps, J. A. (1995). Twins with schizophrenia: Genes or germs. *Schizophrenia Bulletin, 21,* 13–18.

DeCasper, A. J., & Spence, M. J. (1986). Prenatal maternal speech influences newborns' perception of speech sounds. *Infant Behavior and Development, 9,* 133–150.

Denenberg, V. H. (1964). Critical periods, stimulus input, and emotional reactivity: A theory of infantile stimulation. *Psychological Review, 71,* 335–351.

Denenberg, V. H. (1969). The effects of early experience. In E. S. E. Hafez (Ed.), *The behaviour of domestic animals* (2nd ed., pp. 95–130). Baltimore: Williams & Wilkins.

Denenberg, V. H., & Rosenberg, K. M. (1967). Nongenetic transmission of information. *Nature, 216,* 549–550.

Dewey, J., & Bentley, A. F. (1949). *Knowing and the known.* Boston: Beacon.

Dewsbury, D. A. (1978). *Comparative animal behavior.* New York: McGraw-Hill.

DiBerardino, M. A. (1988). Genomic multipotentiality of differentiated somatic cells. In G. Eguchi, T. S. Okada, & L. Saxén, (Eds.), *Regulatory mechanisms in developmental processes* (pp. 129–136). Ireland: Elsevier.

Dmitrieva, L. P., & Gottlieb, G. (1994). Influence of auditory experience on the development of brain stem auditory-evoked potentials in mallard duck embryos and hatchlings. *Behavioral and Neural Biology, 61,* 19–28.

Driesch, H. (1929). *The science and philosophy of the organism* (2nd ed., abridged). London: A. & C. Black. (Original work published 1908).

Dyer, A. B., Lickliter, R. & Gottlieb, G. (1989). Maternal and peer imprinting in mallard ducklings under experimentally simulated natural social conditions. *Developmental Psychobiology, 22,* 463–475.

Edelman, G. M. (1987). *Neural Darwinism: The theory of neuronal group selection.* New York: Basic Books.

Edelman, G. M. (1988). *Topobiology: An introduction to molecular biology.* New York: Basic Books.

Edelman, G. (1992). *Bright air, brilliant fire: On the matter of mind.* New York: Basic Books.

Edelman, G. M., & Jones, F. S. (1993). Outside and downstream of the homeobox. *Journal of Biological Chemistry, 268,* 20683–20686.

Elman, J. L., Bates, E. A., Johnson, M. H., Karmiloff-Smith, A., Parisi, D., & Plunkett, K. (1996). *Rethinking innateness: A connectionist perspective on development.* Cambridge, MA: MIT Press.

Fischer, K. W., (1980). A theory of cognitive development: The control and construction of hierarchies of skill. *Psychological Review, 87,* 477–531.

Fischer, K. W., Bullock, D. H., Rotenberg, E. J., & Raya, P. (1993). The dynamics of competence: How context contributes directly to skill. In R. H. Wozniak and K. W. Fischer (Eds.), *Development in context* (pp. 93–120). Hillsdale, NJ: Lawrence Erlbaum Associates.

Fishbein, H. D.(1976). *Evolution, development, and children's learning.* Pacific Palisades, CA: Goodyear.

Ford, D. H., & Lerner, R. M. (1992). *Developmental systems theory: An integrative approach.* Newbury Park, CA: Sage.

Forgays, D. G., & Forgays, J. W. (1952). The nature of the effect of free-environmental experience in the rat. *Journal of Comparative and Physiological Psychology, 45*, 322–328.

Futuyma, D. J. (1988). *Sturm und Drang* and the evolutionary synthesis. *Evolution, 42*, 217–226.

Gariépy, J.-L. (1995). The evolution of a developmental science: Early determinism, modern interactionism, and a new systemic approach. *Annals of Child Development, 11*, 167–224.

Garstang, W. (1922). The theory of recapitulation: A critical re-statement of the biogenetic law. *Journal of the Linnean Society of London, Zoology, 35*, 81–101.

Gehring, W. J. (1987). Homeoboxes in the study of development. *Science, 236*, 1245–1252.

Gilbert, S. F., Opitz, J. M., & Raff, R. A. (1996). Resynthesizing evolutionary and developmental biology. *Developmental Biology, 173*, 357–372.

Ginty, D. D., Bading, H., & Greenberg, M. E. (1992). Trans-synaptic regulation of gene expression. *Current Opinion in Neurobiology, 2*, 312–316.

Glickman, S. E., & Sroges, R. W. (1966). Curiosity in zoo animals. *Behaviour, 26*, 151–188.

Goldin-Meadow, S. (in press). The resilience of language in humans. In C. T. Snowden & M. Hausberger (Eds.), *Social influences on vocal development.* New York: Cambridge University Press.

Goldschmidt, R. (1933). Some aspects of evolution. *Science, 78*, 539–547.

Goldschmidt, R. (1952). Evolution, as viewed by one geneticist. *American Scientist, 40*, 84–98, 135.

Goodwin, B. C. (1984). A relational or field theory of reproduction and its evolutionary implications. In M.-W. Ho & P. T. Saunders (Eds.), *Beyond neo-Darwinism: An introduction to the new evolutionary paradigm* (pp. 219–242). London: Academic Press.

Gorbman, A., Dickhoff, W. W., Vigna, S. R., Clark, N. B., & Ralph, C. L. (1983). *Comparative endocrinology.* New York: Wiley.

Gottesman, I. I., & Shields, J. (1982). *Schizophrenia: The epigenetic puzzle.* Cambridge, England: Cambridge University Press.

Gottlieb, G. (1965). Imprinting in relation to parental and species identification by avian neonates. *Journal of Comparative and Physiological Psychology, 59*, 345–356.

Gottlieb, G. (1966). Species identification by avian neonates: Contributory effect of perinatal auditory stimulation. *Animal Behaviour, 14*, 282–290.

Gottlieb, G. (1970). Conceptions of prenatal behavior. In L. R. Aronson, E. Tobach, D. S. Lehrman, and J. S. Rosenblatt (Eds.), *Development and evolution of behavior: Essays in memory of T. C. Schneirla* (pp. 111–137). San Francisco: W. H. Freeman.

Gottlieb, G. (1971a). *Development of species identification in birds: An inquiry into the prenatal determinants of perception.* Chicago: University of Chicago Press.

Gottlieb, G. (1971b). Ontogenesis of sensory function in birds and mammals. In E. Tobach, L. R. Aronson, & E. Shaw (Eds.), *The biopsychology of development* (pp. 67–128). New York: Academic Press.

Gottlieb, G. (1974). On the acoustic basis of species identification in wood ducklings. *Journal of Comparative and Physiological Psychology, 87*, 1038–1048.

Gottlieb, G. (1975a). Development of species identification in ducklings: I. Nature of perceptual deficit caused by embryonic auditory deprivation. *Journal of Comparative and Physiological Psychology, 89*, 387–399.

Gottlieb, G. (1975b). Development of species identification in ducklings: II. Experiential prevention of perceptual deficit caused by embryonic auditory deprivation. *Journal of*

Comparative and Physiological Psychology, 89, 675–684.

Gottlieb, G. (1975c). Development of species identification in ducklings: III. Maturational rectification of perceptual deficit caused by embryonic auditory deprivation. *Journal of Comparative and Physiological Psychology, 89,* 899–912.

Gottlieb, G. (1976a). Conceptions of prenatal development: Behavioral embryology. *Psychological Review, 83,* 215–234.

Gottlieb, G. (1976b). The roles of experience in the development of behavior and the nervous system. In G. Gottlieb (Ed.), *Neural and behavioral specificity* (pp. 25–54). New York: Academic Press.

Gottlieb, G. (1978). Development of species identification in ducklings: IV. Change in species-specific perception caused by auditory deprivation. *Journal of Comparative and Physiological Psychology, 92,* 375–387.

Gottlieb, G. (1979). Development of species identification in ducklings: V. Perceptual differentiation in the embryo. *Journal of Comparative and Physiological Psychology, 93,* 831–854.

Gottlieb, G. (1980a). Development of species identification in ducklings: VI. Specific embryonic experience required to maintain species-typical perception in Peking ducklings. *Journal of Comparative and Physiological Psychology, 94,* 579–587.

Gottlieb, G. (1980b). Development of species identification in ducklings: VII. Highly specific early experience fosters species-specific perception in wood ducklings. *Journal of Comparative and Physiological Psychology, 94,* 1019–1027.

Gottlieb, G. (1981). Development of species identification in ducklings: VIII. Embryonic vs. postnatal critical period for the maintenance of species-typical perception. *Journal of Comparative and Physiological Psychology, 95,* 540–547.

Gottlieb, G. (1982). Development of species identification in ducklings: IX. The necessity of experiencing normal variations in embryonic auditory stimulation. *Developmental Psychobiology, 15,* 507–517.

Gottlieb, G. (1983). Development of species identification in ducklings: X. Perceptual specificity in the wood duck embryo requires sib stimulation for maintenance. *Developmental Psychobiology, 16,* 323–333.

Gottlieb, G. (1984). Development of species identification in ducklings: XII. Ineffectiveness of auditory self-stimulation in wood ducklings (*Aix sponsa*). *Journal of Comparative Psychology, 98,* 137–141.

Gottlieb, G. (1985). Development of species identification in ducklings: XI. Embryonic critical period for species-typical perception in the hatchling. *Animal Behaviour, 33,* 225–233.

Gottlieb, G. (1987a). A tribute to Clarence Luther Herrick (1858–1904): Founder of developmental psychobiology. *Developmental Psychobiology, 20,* 1–5.

Gottlieb, G. (1987b). Development of species identification in ducklings: XIII. A comparison of malleable and critical periods of perceptual development. *Developmental Psychobiology, 20,* 393–404.

Gottlieb, G. (1987c). Development of species identification in ducklings: XIV. Malleability of species-specific perception. *Journal of Comparative Psychology, 101,* 178–182.

Gottlieb, G. (1991a). Experiential canalization of behavioral development: Theory. *Developmental Psychology, 27,* 4–13.

Gottlieb, G. (1991b). Social induction of malleability in ducklings. *Animal Behaviour, 41,* 953–962.

Gottlieb, G. (1991c). Experiential canalization of behavioral development: Results. *Developmental Psychology, 27,* 35–39.

Gottlieb, G. (1992). *Individual development and evolution.* New York: Oxford University Press.

Gottlieb, G. (1993). Social induction of malleability in ducklings: Sensory basis and psychological mechanism. *Animal Behaviour, 45,* 707–719.

Gottlieb, G. (1995). Some conceptual deficiencies in "developmental" behavior genetics. *Human Development, 38,* 131-169.

Gottlieb, G. (1996). A systems view of psychobiological development. In D. Magnusson (Ed.), *The lifespan development of individuals: Behavioral, neurobiological, and psychosocial perspectives* (pp. 76-103). Cambridge, England: Cambridge University Press.

Gottlieb, G., Johnston, T. D., & Scoville, R. P. (1982). Conceptions of development and the evolution of behavior. *Behavioral and Brain Sciences, 2,* 284.

Gottlieb, G., & Kuo, Z.-Y. (1965). Development of behavior in the duck embryo. *Journal of Comparative and Physiological Psychology, 59,* 183-188.

Gottlieb, G., Tomlinson, W. T., & Radell, P. (1989). Developmental intersensory interference: Premature visual experience suppresses auditory learning in ducklings. *Infant Behavior and Development, 12,* 1-12.

Gottlieb, G., & Vandenbergh, J. G. (1968). Ontogeny of vocalization in duck and chick embryos. *Journal of Experimental Zoology, 168,* 307-325.

Gottlieb, G., Wahlsten, D., & Lickliter, R. (in press). The significance of biology for human development: A developmental psychobiological systems view. In R. M. Lerner (Ed.), *Handbook of child psychology: Vol. 1. Theoretical models of human development* (5th ed.). New York: Wiley.

Gray, L. (1990). Activity level and auditory responsiveness in neonatal chickens. *Developmental Psychobiology, 23,* 297-308.

Gray, R. (1992). Death of the gene: Developmental systems strike back. In P. Griffiths (Ed.), *Trees of life: Essays on the philosophy of biology* (pp. 165-209). Dordrecht: Kluwer.

Greenough, W. T., & Juraska, J. M. (1979). Experience-induced changes in brain fine structure: Their behavioral implications. In M. E. Hahn, C. Jensen, and B. C. Dudek (Eds.), *Development and evolution of brain size* (pp. 296-320). New York: Academic Press.

Grene, M. (1987). Hierarchies in biology. *American Scientist, 75,* 504-510.

Griffiths, P. E., & Gray, R. D. (1994). Developmental systems and evolutionary explanation. *Journal of Philosophy, 91,* 277-304.

Grouse, L. D., Schrier, B. K., Letendre, C. H., & Nelson, P. G. (1980). RNA sequence complexity in central nervous system development and plasticity. *Current Topics in Developmental Biology, 16,* 381-397.

Guyomarc'h, J.-C. (1972) Les bases ontogénétiques de l'attractivité du gloussement maternel chez la poule domestique. *Revue du Comportement Animal, 6,* 79-94.

Hall, B. K. (1988). The embryonic development of bone. *American Scientist, 76,* 174-181.

Hamburger, V. (1957). The concept of "development" in biology. In D. H. Harris (Ed.), *The concept of development* (pp. 49-58). Minneapolis: University of Minnesota.

Hamburger, V. (1964). Ontogeny of behaviour and its structural basis. In D. Richter (Ed.), *Comparative neurochemistry* (pp. 21-34). Oxford: Pergamon Press.

Hamburger, V. (1973). Anatomical and physiological basis of embryonic motility in birds and mammals. In G. Gottlieb (Ed.), *Behavioral embryology* (pp. 51-76). New York: Academic Press.

Hamburger, V. (1988). *The heritage of experimental embryology: Hans Spemann and the organizer.* New York: Oxford University Press.

Hardy, A. C. (1965). *The living stream.* London: Collins.

Harlow, H. F. (1958). The nature of love. *American Psychologist, 13,* 673-685.

Harlow, H. F., Dodsworth, R. O., & Harlow, M. K. (1965). Total social isolation in monkeys. *Proceedings of the National Academy of Sciences USA, 54,* 90-96.

Harter, M. R. (1991). Event-related potential indices: learning disabilities and visual processing. In J. E. Obrzut & G. W. Hynd (Eds.), *Neuropsychological foundations of learning disabilites* (pp. 437-473). San Diego: Academic Press.

Heaton, M. B. (1971). *Stimulus coding in the species-specific perception of Peking ducklings.*

Unpublished doctoral dissertation, North Carolina State University.

Hebb, D. O. (1955). Drive and the C.N.S. (Conceptual Nervous System). *Psychological Review, 62,* 243-254.

Held, R., & Hein, A. (1963). Movement-produced stimulation in the development of visually guided behavior. *Journal of Comparative and Physiological Psychology, 81,* 394-398.

Hess, E. H. (1958). "Imprinting" in animals. *Scientific American, 198,* 81-90.

Hinde, R. A. (1990). The interdependence of the behavioral sciences. *Philosophical Transactions of the Royal Society London B., 329,* 217-227.

His, W. (1888). On the principles of animal morphology. *Proceedings of the Royal Society of Edinburgh, 15,* 287-298.

Ho, M.-W. (1984). Environment and heredity in development and evolution. In M.-W. Ho & P. T. Saunders (Eds.), *Beyond neo-Darwinism: An introduction to the new evolutionary paradigm* (pp. 267-289). London: Academic Press.

Ho, M.-W., & Saunders, P. T. (1982). The epigenetic approach to the evolution of organisms—With notes on its relevance to social and cultural evolution. In H. C. Plotkin (Ed.), *Learning, development, and culture: Essays in evolutionary epistemology* (pp. 343-361). London: Wiley.

Holliday, R. (1990). Mechanisms for the control of gene activity during development. *Biological Reviews, 65,* 431-471.

Holt, E. B. (1931). *Animal drive and the learning process* (Vol. 1). New York: Holt.

Honey, R. C. (1990). Stimulus generalization as a function of stimulus novelty and familiarity in rats. *Journal of Experimental Psychology: Animal Behavior Processes, 16,* 178-184.

Horowitz, F. D. (1987). *Exploring developmental theories: Toward a structural/behavioral model of development.* Hillsdale, NJ: Lawrence Erlbaum Associates.

Horton, J. C., & Hocking, D. R. (1996). An adult-like pattern of ocular dominance columns in striate cortex of newborn monkeys prior to visual experience. *Journal of Neuroscience, 16,* 1791-1807.

Huxley, J. S. (1957). The three types of evolutionary progress. *Nature, 180,* 454-455.

Hydén, H., & Egyházi, E. (1962). Nuclear RNA changes of nerve cells during a learning experiment in rats. *Proceedings of the National Academy of Sciences USA, 48,* 1366-1373.

Hydén, H., & Egyházi, E. (1964). Changes in RNA content and base composition in cortical neurons of rats in a learning experiment involving transfer of handedness. *Proceedings of the National Academy of Sciences USA, 52,* 1030-1035.

Hymovitch, B. (1952). The effects of experimental variations on problem solving in the rat. *Journal of Comparative and Physiological Psychology, 45,* 313-321.

Igel, G. J., & Calvin, A. D. (1960). The development of affectional responses in infant dogs. *Journal of Comparative and Physiological Psychology, 53,* 302-305.

Ingham, P. W. (1988). The molecular genetics of embryonic pattern formation in *Drosophila. Nature, 335,* 25-34.

Jaisson, P. (1975). L'impregnation dans l'ontogenese des comportements de soins aux cocons chez la jeune formi rousse (*Formica polyctena* Forst). *Behaviour, 52,* 1-37.

Jeddi, E. (1970). Comfort de contact et thermoregulation comportementale. *Physiology and Behavior, 5,* 1487-1493.

Jerison, H. J. (1969). Brain evolution and dinosaur brains. *American Naturalist, 103,* 575-588.

Jerison, H. J. (1973). *Evolution of the brain and intelligence.* New York: Academic Press.

John, B., & Miklos, G. L. G. (1988). *The eukaryote genome in development and evolution.* London: Allen & Unwin.

Johnston, T. D. (1987). The persistence of dichotomies in the study of behavioral development. *Developmental Review, 7,* 149-182.

Johnston, T. D., & Gottlieb, G. (1990). Neophenogenesis: A developmental theory of phenotypic evolution. *Journal of Theoretical Biology, 147,* 471-495.

Johnston, T. D., & Hyatt, L. E. (1994). *Genes, interactions, and the development of behavior.*

Unpublished manuscript.

Klopfer, P. H. (1959). An analysis of learning in young Anatidae. *Ecology, 40*, 90–102.

Kollar, E. J., & Fisher, C. (1980). Tooth induction in chick epithelium: Expression of quiescent genes for enamel synthesis. *Science, 207*, 993–995.

Konishi, M. (1963). The role of auditory feedback in the vocal behavior of the domestic fowl. *Zeitschrift für Tierpsychologie, 20*, 349–367.

Kovach, J. K., & Wilson, G. (1988). Genetics of color preferences in quail chicks: Major genes and variable buffering by background genotype. *Behavior Genetics, 18*, 645–661.

Kroodsma, D. E., & Konishi, M. (1991). A suboscine bird (eastern phoebe, *Sayornis phoebe)* develops normal song without auditory feedback. *Animal Behaviour, 42*, 477–488.

Kroodsma, D. E., & Miller, E. H. (Eds.). (1982). *Acoustic communication in birds, Vol. 2, Song learning and its consequences.* New York: Academic Press.

Kuhn, D. (1995). Microgenetic study of change: What has it told us? *Psychological Science, 3*, 133–139.

Kuo, Z.-Y. (1921). Giving up instincts in psychology. *Journal of Philosophy, 18*, 645–664.

Kuo, Z.-Y. (1967). *The dynamics of behavior development.* New York: Random House.

Kuo, Z.-Y. (1976). *The dynamics of behavior development* (enlarged ed.). New York: Plenum.

Lamarck, J. B. (1984). *Zoological pholosophy: An exposition with regard to the natural history of animals.* Chicago: University of Chicago Press. (Original work published in French, 1809)

Larson, A., Prager, E. M., & Wilson, A. C. (1984). Chromosomal evolution, speciation and morphological change in vertebrates: The role of social behavior. *Chromosomes Today, 8*, 215–228.

Lee, C., Parikh, V., Itsukaichi, T., Bae, K., & Edery, I. (1996). Resetting the *Drosophila* clock by photic regulation of PER and a PER-TIM complex. *Science, 271*, 740–744.

Lehrman, D. S. (1953). A critique of Konrad Lorenz's theory of instinctive behavior. *Quarterly Review of Biology, 28*, 337–363.

Lehrman, D. S. (1970). Semantic and conceptual issues in the nature-nurture problem. In L. R. Aronson, E. Tobach, D. S. Lehrman, & J. S. Rosenblatt (Eds.), *Development and evolution of behavior: Essays in memory of T. C. Schneirla* (pp. 17–52). San Francisco: W. H. Freeman.

Leonovicová, V., & Novák, V. J. A. (Eds.). (1987). *Behavior as one of the main factors of evolution.* Pvaha, Czechoslovakia: Czechoslovak Academy of Sciences.

Lerner, R. M., & Kauffman, M. B. (1985). The concept of development in contextualism. *Developmental Review, 5*, 309–333.

Levine, S. (1962). The effects of infantile experience on adult behavior. In A. J. Bachrach (Ed.), *Experimental foundations of clinical psychology* (pp. 139–169). New York: Basic Books.

Lickliter, R., & Banker, H. (1994). Prenatal components of intersensory development in precocial birds. In D. J. Lewkowicz & R. Lickliter (Eds.), *The development of intersensory perception: Comparative aspects* (pp. 56–80). Hillsdale, NJ: Lawrence Erlbaum Associates.

Locke, J. L. (1993). *The child's path to spoken language.* Cambridge, MA: Harvard University Press.

Logan, C. A. (1985). Development as explanation. *Applied Developmental Psychology, 2*, 1–32.

Logan, C. A. (1992). Developmental analysis in behavioral systems: The case of bird song. *Annals of the New York Academy of Sciences, 662*, 102–117.

London, I. D. (1949). The concept of the behavioral spectrum. *Journal of Genetic Psychology, 24*, 177–184.

Lorenz, K. (1937). The companion in the bird's world. *Auk, 54*, 245–273.

Lorenz, K. (1965). *Evolution and modification of behavior.* Chicago: University of Chicago

Press.

Løvtrup, S. (1987). *Darwinism: Refutation of a myth*. Beckenham, England: Croom Helm.

Lucero, M. A. (1970). Lengthening of REM sleep duration consequent to learning in the rat. *Brain Research, 20*, 319–322.

Lumsden, C. J., & Wilson, E. O. (1980). Translation of epigenetic rules of individual behavior into ethnographic patterns. *Proceedings of the National Academy of Sciences USA, 77*, 4382–4386.

Mack, K. J., & Mack, P. A. (1992). Induction of transcription factors in somatosensory cortex after tactile stimulation. *Molecular Brain Research, 12*, 141–147.

Macphail, E. M., & Reilly, S. (1989). Rapid acquisition of a novelty versus familiarity concept by pigeons (*Columba livia*). *Journal of Experimental Psychology: Animal Behavior Processes, 15*, 242–252.

Magnusson, D., & Törestad, B. (1993). A holistic view of personality: A model revisited. *Annual Review of Psychology, 44*, 427–452.

Mangold, O., & Seidel, F. (1927). Homoplastische und heteroplastische Verschmelzung ganzer Tritonkeime. *Roux's Archiv für Entwicklungsmechanik der Organismen, 111*, 593–665.

Marler, P., Zoloth, S., & Dooling, R. (1981). Innate programs for perceptual development: An ethological view. In E. Gollin (Ed.), *Developmental plasticity* (pp. 135–172). New York: Academic Press.

Masataka, N. (1994). Effects of experience with live insects on the development of fear of snakes in squirrel monkeys, *Saimiri* sciurens. *Animal Behaviour, 46*, 741–746.

Mason, W. A. (1968). Early social deprivation in the nonhuman primates: Implications for human behavior. In D. Glass (Ed.), *Biology and behavior: Environmental influences* (pp. 70–101). New York: Rockefeller University Press.

Mauro, V. P., Wood, I. C., Krushel, L., Crossin, K. L., & Edelman, G. M. (1994). Cell adhesion alters gene transcription in chicken embryo brain cells and mouse embryonal carcinoma cells. *Proceedings of the National Academy of Sciences USA, 91*, 2868–2872.

Mayr, E. (1954). Change of genetic environment and evolution. In J. Huxley, A. C. Hardy, & E. B. Ford (Eds.), *Evolution as a process* (pp. 157–180). London: Allen and Unwin.

Mayr, E. (1982). *The growth of biological thought*. Cambridge, MA: Harvard University Press.

Mayr, E. (1988). *Toward a new philosophy of biology: Observations of an evolutionist*. Cambridge, MA: Belknap Press of Harvard University Press.

McDonald, P., & Topoff, H. (1985). Social regulation of behavioral development in the ant, *Novomessor albisetosus* (Mayr). *Journal of Comparative Psychology, 99*, 3–14.

Michel, G. F., & Moore, C. L. (1995). *Developmental psychobiology: An interdisciplinary science*. Cambridge, MA: MIT Press.

Miklos, G. L. G. (1993). Molecules and cognition: The latterday lessons of levels, language, and lac. Evolutionary overview of brain structure and function in some vertebrates and invertebrates. *Journal of Neurobiology, 24*, 842–890.

Miklos, G. L. G., & Edelman, G. M. (1996). *Complexity in gene function and morphogenesis: Major challenges of the post-sequence era*. Unpublished manuscript.

Miller, D. B. (1988). Development of instinctive behavior: An epigenetic and ecological approach. In E. M. Blass (Ed.), *Handbook of behavioral neurobiology, Vol. 9: Developmental psychology and behavioral ecology* (pp. 415–444). New York: Plenum Press.

Miller, D. B. (1994). Social context affects the ontogeny of instinctive behaviour. *Animal Behaviour, 48*, 627–634.

Miller, D. B. (in press). The effects of nonobvious forms of experience on the development of instinctive behavior. In C. Dent-Read & P. Zukow-Goldring (Eds.), *Changing ecological approaches to development: Organism-environment mutualities*. Washington, DC: American Psychological Association.

Miller, D. B., & Gottlieb, G. (1976). Acoustic features of wood duck (*Aix sponsa*) maternal calls. *Behaviour, 57,* 260–280.

Miller, D. B., & Gottlieb, G. (1978). Maternal vocalizations of mallard ducks (*Anas platyrhynchos*). *Animal Behaviour, 26,* 1178–1194.

Miller, D. B., & Gottlieb, G. (1981). Effects of domestication on production and perception of mallard maternal alarm calls: Developmental lag in behavioral arousal. *Journal of Comparative and Physiological Psychology, 95,* 205–219.

Miller, D. B., Hicinbothom, G., & Blaich, C. F. (1990). Alarm call responsivity of mallard ducklings: Multiple pathways in behavioural development. *Animal Behaviour, 39,* 1207–1212.

Miller, P. H. (1989). *Theories of developmental psychology* (2nd ed). San Francisco: Freeman.

Mirsky, A. E., & Ris, H. (1951). The deoxyribonucleic acid content of animal cells and its evolutionary significance. *Journal of General Physiology 34,* 451–462.

Mivart, St. G. (1871). *On the genesis of species.* London: Macmillan.

Montagu, A. (1945). Intelligence of northern negroes and southern whites in the First World War. *American Journal of Psychology, 58,* 451–462.

Montagu, A. (1977). Sociogenic brain damage. In S. Arieti and G. Chrzanowski (Eds.), *New dimensions in psychiatry: A world view* (Vol. 2, pp. 4–25). New York: Wiley.

Morgan, J. I., & Curran, T. (1991). Stimulus-transcription coupling in the nervous system: Involvement of the inducible proto-oncogenes fos and jun. *Annual Review of Neuroscience, 14,* 421–451.

Müller, A. (1991). *Interaction and determination.* Budapest: Adadémai Kiadó.

Myers, M. M., Brunelli, S. A., Shair, H. M., Squire, J. M., & Hofer, M. A. (1989). Relationships between maternal behavior of SHR and WKY dams and adult blood pressures of cross-fostered F_1 pups. *Developmental Psychobiology, 22,* 55–67.

Myers, M. M., Brunelli, S. A., Squire, J. M., Shindeldecker, R. D., & Hofer, M. A. (1989). Maternal behavior of SHR rats and its relationships to offspring blood pressure. *Developmental Psychobiology, 22,* 29–53.

Myers, M. P., Wager-Smith, K., Rothenfluh-Hilfiker, A., & Young, M. W. (1996). Light-induced degradation of TIMELESS and entrainment of the *Drosophila* circadian clock. *Science, 271,* 736–740.

Needham, J. (1959). *A history of embryology.* New York: Abelard-Schuman.

Nöel, M. (1989). Early development in mice. V. Sensorimotor development of four coisogenic mutant strains. *Physiology and Behavior, 45,* 21–26.

Nottebohm, F., & Nottebohm, M. (1971). Vocalizations and breeding behaviour of surgically deafened ring doves (*Streptopelia risoria*). *Animal Behaviour, 19,* 313–327.

Oppenheim, R. W. (1974). The ontogeny of behavior in the chick embryo. *Advances in the Study of Behavior, 5,* 133–172.

Oyama, S. (1985). *The ontogeny of information.* Cambridge, England: Cambridge University Press.

Panksepp, J., Bean, N. J., Bishop, P., Vilberg, T., & Sahley, T. I. (1980). Opioid blockade and social comfort in chicks. *Pharmacology, Biochemistry & Behavior, 13,* 673–683.

Parker, S. T., & Gibson, K. R. (1979). A developmental model for the evolution of language and intelligence in early hominids. *Behavioral and Brain Sciences, 2,* 367–408.

Parsons, P. A. (1981). Habitat selection and speciation in *Drosophila.* In W. R. Atchley & D. S. Woodruff (Eds.), *Evolution and speciation: Essays in honor of M. J. D. White* (pp. 219–240). Cambridge, England: Cambridge University Press.

Perry, H. S. (1982). *Psychiatrist of America: The life of Harry Stack Sullivan.* Cambridge, MA: Belknap Press of Harvard University Press.

Piaget, J. (1978). *Behavior and evolution.* New York: Pantheon Books.

Platt, S. A., & Sanislow, C. A. (1988). Norm-of-reaction: Definition and misinterpretation of animal research. *Journal of Comparative Psychology, 102,* 254–261.

Plotkin, H. C. (Ed.). (1988). *Th role of behavior in evolution.* Cambridge, MA: MIT Press.

Pritchard, D. J. (1986). *Foundations of developmental genetics*. London: Taylor and Francis.

Raff, R. A., & Kaufman, T. C. (1983). *Embroyos, genes, and evolution*. New York: Macmillan.

Razran, G. (1971). *Mind in evolution*. New York: Houghton Mifflin.

Reid, R. G. B. (1985). *Evolutionary theory: The unfinished synthesis*. Ithaca, NY: Cornell University Press.

Renner, M. J., & Rosenzweig, M. R. (1987). *Enriched and impoverished environments*. New York: Springer.

Rensch, B. (1959). *Evolution above the species level*. New York: Columbia University Press.

Rosen, D. E., & Buth, D. G. (1980). Empirical evolutionary research versus neo-Darwinian speculation. *Systematic Zoology, 29*, 300–308.

Rosen, K. M., McCormack, M. A., Villa-Komaroff, L., & Mower, G. D. (1992). Brief visual experience induces immediate early gene expression in the cat visual cortex. *Proceedings of the National Academy of Sciences USA, 89*, 5437–5441.

Roubertoux, P. L., Nosten-Bertrand, M., & Carlier, M. (1990). Additive and interactive effects of genotype and maternal environment. *Advances in the Study of Behavior, 19*, 205–247.

Roux, W. (1974). Contributions to the developmental mechanics of the embryo. In B. H. Willier & J. M. Oppenheimer (Eds.), *Foundations of experimental embryology* (pp. 2–37). New York: Hafner. (Original work published in German, 1888)

Rubel, E. W , & Parks, T. N. (1988). Organization and development of the avian brain-stem auditory system. In G. M. Edelman, W. E. Gall, & W. M. Cowan (Eds.), *Auditory function: Neurobiological bases of hearing (pp. 3–92). New York: Wiley.*

Rustak, B., Robertson, H. A., Wisden, W., & Hunt, S. P. (1990). Light pulses that shift rhythms induce gene expression in the suprachiasmatic nucleus. *Science, 248*, 1237–1240.

Sackett, G. P. (1968). Abnormal behavior in laboratory-reared rhesus monkeys. In M. W. Fox (Ed.), *Abnormal behavior in animals* (pp. 293–331). Philadelphia: W. B. Saunders.

Saint-Hilaire, E. G. (1825). Sur les déviations organiques provoquées et observées dans un éstablissement des incubations artificielles. *Mémoires. Museum National d'Histoire Naturelle (Paris), 13*, 289–296.

Salthe, S. N. (1985). *Evolving hierarchical systems: Their structure and representation*. New York: Columbia University Press.

Sameroff, A. J. (1983). Developmental systems: Contexts and evolution. In P. H. Mussen (Series Ed.) & W. Kessen (Vol. Ed.), *Handbook of child psychology, Vol. 1: History, theory, and methods* (pp. 237–294). New York: Wiley.

de Santillana, G., & von Dechend, H. (1977). Preface. *Hamlet's mill* (vii–xiii). Boston: David R. Godine.

Sarich, V. (1980). A macromolecular perspective on the material basis of evolution. In L. K. Piternick (Ed.), *Richard Goldschmidt: Controversial geneticist and creative biologist* (pp. 27–31). Boston: Birkhäuser.

Scarr, S. (1993). Biological and cultural diversity: The legacy of Darwin for development. *Child Development, 64*, 1333–1353.

Scarr, S. (1997). Behavior genetics and socialization theories of intelligence: Truce and reconciliation. In R. J. Sternberg & E. Grigorenko (Eds.), *Intelligence: Heredity and environment* (pp. 3–41). New York: Cambridge University Press.

Scarr-Salapatek, S. (1976). Genetic determinants of infant development: An overstated case. In L. Lipsitt (Ed.), *Developmental psychobiology: The significance of infancy* (pp. 59–79). Hillsdale, NJ: Lawrence Erlbaum Associates.

Schneirla, T. C. (1956). Interrelationships of the "innate" and the "acquired" in instinctive behavior. In P.-P. Grassé (Ed.), *L'Instinct dans le comportement des animaux et de l'homme* (pp. 387–452). Paris: Masson.

Schneirla, T. C. (1960). Instinctive behavior, maturation—Experience and development. In B.

Kaplan & S. Wapner, (Eds.), *Perspectives in psychological theory—Essays in honor of Heinz Werner* (pp. 303–334). New York: International Universities Press.

Schneirla, T. C. (1965). Aspects of stimulation and organization in approach/withdrawal processes underlying vertebrate behavioral development. *Advances in the Study of Behavior*, 1, 1–71.

Scoville, R. P. (1982). *Embryonic development of neonatal vocalizations in Peking ducklings (Anas platyrhynchos).* Unpublished doctoral dissertation, University of North Carolina at Chapel Hill.

Scoville, R. P., & Gottlieb, G. (1978). The calculation of repetition rate in avian vocalizations. *Animal Behaviour*, 26, 962–963.

Scoville, R. S., & Gottlieb, G. (1980). Development of vocal behavior in Peking ducklings. *Animal Behaviour*, 28, 1095–1109.

Sewertzoff, A. N. (1929). Directions of evolution. *Acta Zoologica* (Stockholm), 10, 59–141.

Sexton, C. A. (1994). *Sociogenesis of species-typical and species-atypical behavior in mallard ducklings.* Unpublished doctoral dissertation, University of North Carolina at Greensboro.

Shanahan, M. J., Valsiner, J., & Gottlieb, G. (1997). Developmental concepts across disciplines. In J. Tudge, M. J. Shanahan, & J. Valsiner (Eds.), *Comparisons in human development* (pp. 34–71). New York: Cambridge University Press.

Shapiro, D. Y. (1981). Serial female sex changes after simultaneous removal of males from social groups of a coral reef fish. *Science*, 209, 1136–1137.

Shatz, C. (1990). Impulse activity and the patterning of connections during CNS development. *Neuron*, 5, 745–756.

Shatz, C. (1994). Role for spontaneous neural activity in the patterning of connections between retina and LGN during visual system development. *International Journal of Developmental Neuroscience*, 12, 531–546.

Sheng, M., & Greenberg, M. E. (1990). The regulation and function of c-fos and other immediate early genes in the nervous system. *Neuron*, 4, 477–485.

Singh, R. S. (1989). Population genetics and the evolution of species related to *Drosophila melanogaster*. *Annual Review of Genetics*, 23, 425–453.

Smuts, J. C. (1926). *Holism and evolution.* London: Macmillan.

Sparrow, A. H., Price, H. J., & Underbrink, A. G. (1972). A survey of DNA content per cell and per chromosome of prokaryotic and eukaryotic organisms: Some evolutionary considerations. *Brookhaven Symposium on Biology*, 23, 451–494.

Spelke, E. S., & Newport, E. L. (in press). Nativism, empiricism, and the development of knowledge. In R. M. Lerner (Ed.), *Handbook of child psychology: Theoretical models of human development* (Vol. 1, 5th ed.). New York: Wiley.

Sperry, R. W. (1951). Mechanisms of neural maturation. In S. S. Stevens (Ed.), *Handbook of experimental psychology* (pp. 236–280). New York: Wiley.

Sperry, R. W. (1971). How a developing brain gets itself properly wired for adaptive function. In E. Tobach, L. R. Aronson, & E. Shaw (Eds.), *The biopsychology of development* (pp. 28–34). New York: Academic Press.

Steele, E. J. (1979). *Somatic selection and adaptive evolution.* Toronto: Williams and Wallace.

Stern, W. (1938). *General psychology from the personalistic standpoint.* New York: Macmillan.

Strohman, R. C. (1993). Book reviews. *Integrative Physiological and Behavioral Science*, 28, 99–110.

Sullivan, H. S. (1953). *The interpersonal theory of psychiatry.* New York: Norton.

Tagney, J. (1973). Sleep patterns related to rearing rats in enriched and impoverished environments. *Brain Research*, 53, 353–361.

Tees, R. C. (1990). Experience, perceptual competencies, and rat cortex. In B. Kolb & R. C. Tees (Ed.), *The cerebral cortex of the rat* (pp. 507–536). Cambridge, MA: MIT Press.

Thelen, E., & Smith, L. B. (1994). *A dynamic systems approach to cognition and action.*

Cambridge, MA: MIT Press.

Thoman, E. B. (1990). Sleeping and waking states in infants: A functional perspective. *Neuroscience and Biobehavioral Reviews, 14*, 93-107.

Thomas, C. A. (1971). The genetic organization of chromosomes. *Annual Review of Genetics, 5*, 237-256.

Thorpe, W. H. (1963). *Learning and instinct in animals* (2nd ed.). Cambridge, MA: Harvard University Press. (Original work published 1956)

Tolman, E. C. (1932). *Purposive behavior in animals and man.* New York: Century.

Turkewitz, G. (1988). A prenatal souce for the development of hemispheric specialization. In D. L. Molfese & S. J. Segalowitz (Eds.), *Brain lateralization in children* (pp. 73-81). New York: Guilford Press.

Turkewitz, G. & Mellon, R. C. (1989). Dynamic organization of intersensory function. *Canadian Journal of Psychology, 43*, 286-307.

Uphouse, L. L., & Bonner, J. (1975). Preliminary evidence for the effects of environmental complexity on hybridization of rat brain RNA to rat unique DNA. *Developmental Psychobiology, 8*, 171-178.

Valsiner, J. (1987). *Culture and the development of children's action.* Chichester, England: Wiley.

van der Veer, R., & Valsiner, J. (1991). *Understanding Vygotsky: A quest for synthesis.* Oxford: Blackwell.

Waddington, C. H. (1942). Canalization of development and the inheritance of acquired characters. *Nature, 150*, 563-564.

Waddington, C. H. (1953). Genetic assimilation of an acquired character. *Evolution, 7*, 118-126.

Waddington, C. H. (1957). *The strategy of the genes.* London: Allen and Unwin.

Waddington, C. H. (1968). The basic ideas of biology. In C. H. Waddington (Ed.), *Towards a theoretical biology. I. Prolegomena* (pp. 1-32). Chicago: Aldine.

Waddington, C. H. (1971). Concepts of development. In E. Tobach, L. R. Aronson, & E. Shaw (Eds.), *The biopsychology of development* (pp. 17-23). New York: Academic Press.

Wahlsten, D., & Gottlieb, G. (1997). The invalid separation of effects of nature and nurture: Lessons from animal experimentation. In R. J. Sternberg & E. Grigorenko (Eds.), *Intelligence: Heredity and environment* (pp. 163-192). New York: Cambridge University Press.

Wallman, J. (1979). A minimal visual restriction experiment: Preventing chicks from seeing their feet affects later responses to mealworms. *Developmental Psychobiology, 12*, 391-397.

van der Weele, C. (1995). *Images of development: Environmental causes in ontogeny.* Unpublished doctoral dissertation, Vrije University, Amsterdam.

Weismann, A. (1894). *The effect of external influences upon development.* London: Henry Frowde.

Weiss, P. (1959). Cellular dynamics. *Reviews of Modern Physics, 31*, 11-20.

Weiss, P. (1969). *Principles of development.* (Original work published 1939) New York: Hafner.

Wessells, N. K. (1977). *Tissue interactions and development.* Menlo Park, CA: W. A. Benjamin.

Werker, J. F., & Tees, R. C. (1984). Cross-language speech perception: Evidence for perceptual reorganization during the first year of life. *Infant Behavior and Development, 7*, 49-63.

West, M. J., King, A. P., & Freeberg, T. M. (1994). The nature and nurture of neo-phenotypes: A case history. In L. A. Peal (Ed.), *Behavioral mechanisms in evolutionary ecology* (pp. 238-257). Chicago: University of Chicago Press.

Whitehead, A. N. (1929). *Process and reality.* Cambridge, England: Cambridge University

Press.

Williams, L. & Golenski, J. (1970). Infant behavioral state and speech sound discrimination. *Child Development, 50,* 1243-1246.

Wright, S. (1968). *Evolution and the genetics of population, Vol. 1: Genetic and biometric foundations.* Chicago: University of Chicago Press.

Wyles, J. S., Kunkel, J. G., & Wilson, A. C. (1983). Birds, behavior, and anatomical evolution. *Proceedings of the National Academy of Sciences USA, 80,* 4394-4397.

Zamenhof, S., & van Marthens, E. (1978). Nutritional influences on prenatal brain development. In G. Gottlieb (Ed.), *Early influences* (pp. 149-186). New York: Academic Press.

Zamenhof, S., & van Marthens, E. (1979). Brain weight, brain chemical content, and their early manipulation. In M. E. Hahn, C. Jensen, & B. C. Dudek (Eds.), *Development and evolution of brain size* (pp. 164-185). New York: Academic Press.

Name Index

Subject Index

SYNTHESIZING *nature* – *nurture*

prenatal roots of instinctive behavior

GILBERT GOTTLIEB

In this new book, Gilbert Gottlieb, a comparative and developmental scientist of unparalleled theoretical sophistication and empirical accomplishment, turns his singularly creative attention to an analysis of the concept of probabilistic epigenesis. The result is a book that makes a unique and invaluable contribution to an integrated understanding of developmental systems, evolution, and plasticity. This superb volume is a watershed event in scholarship pertinent to the biological and contextual bases of development and is a necessary resource for all scholars concerned with the process of development.

—Dr. Richard M. Lerner
Anita L. Brennan Professor of Education and Director,
Center for Child, Family, and Community Partnerships
Boston College

In contrast to the current consensus that believes biological science has nothing more to learn, Gottlieb's astonishing new book opens wide the door to a new "developmental" view of the relationship between genes and early experiences, opening up the evolution of the mind—in the womb and during childhood—to the possibility of greatly improving what we mislabel "human nature."

—Lloyd deMause, Editor
The Journal of Psychohistory and
Director of The Institute for Psychohistory

ISBN 0-8058-2870-2

90000

9 780805 828702
ISBN 0-8058-2870-2